Hands-on Science

36 fun science activities

- Motivate
- Experience
- Explain
- Apply
- Review and reflect

6516C

HANDS-ON SCIENCE *(Upper)*

Published by Prim-Ed Publishing 2008
Reprinted under licence by Prim-Ed Publishing 2008
Copyright© R.I.C. Publications® 2007
ISBN 978-1-84654-144-5
PR–6516

Additional titles available in this series:
HANDS-ON SCIENCE *(Lower)*
HANDS-ON SCIENCE *(Middle)*

This master may only be reproduced by the original purchaser for use with their class(es). The publisher prohibits the loaning or onselling of this master for the purposes of reproduction.

Copyright Notice

Blackline masters or copy masters are published and sold with a limited copyright. This copyright allows publishers to provide teachers and schools with a wide range of learning activities without copyright being breached. This limited copyright allows the purchaser to make sufficient copies for use within their own education institution. The copyright is not transferable, nor can it be onsold. Following these instructions is not essential but will ensure that you, as the purchaser, have evidence of legal ownership to the copyright if inspection occurs.

For your added protection in the case of copyright inspection, please complete the form below. Retain this form, the complete original document and the invoice or receipt as proof of purchase.

Name of Purchaser:

_____

Date of Purchase:

_____

Supplier:

_____

School Order# (if applicable):

_____

Signature of Purchaser:

_____

Internet websites
In some cases, websites or specific URLs may be recommended. While these are checked and rechecked at the time of publication, the publisher has no control over any subsequent changes which may be made to webpages. It is *strongly* recommended that the class teacher checks *all* URLs before allowing pupils to access them.

View all pages online   **Website:** www.prim-ed.com

Foreword

Hands-on learning is 'learning by doing'. It requires pupils to become active participants as they investigate, experiment, design, create, role-play, cook and more, gaining an understanding of essential scientific concepts from these experiences.

Hands-on learning motivates pupils and engages them in their learning. Instead of being told 'why' something occurs, they see it for themselves, directly observing science in action.

The fun, pupil-orientated activities in the *Hands-on science* series teach scientific concepts and skills, while promoting pupil participation, enthusiasm and curiosity about science. Easily integrated into any primary science programme, *Hands-on science* provides clear, step-by-step instructions for each activity and comprehensive background information for the teacher. A glossary of scientific terms used is also included.

Hands-on science provides pupils with the opportunity to enhance their knowledge of the world around them and to engage in collaborative, fun learning that makes science interesting and exciting!

This book is also provided in digital format on the accompanying CD.

Titles in this series are: *Hands-on Science Lower*
Hands-on Science Middle
Hands-on Science Upper

Contents

Teacher information	**iv – vii**
Why 'hands on'?	vi
Safety	vi
Assessment	vii
Curriculum links	**vii–ix**
Frameworks	**x – xvii**
Assessment	x
Science reflection	xi
Science report	xii
Science recount	xiii
Science investigation	xiv
Science journal	xv
Scientific diagram	xvi
Before and after	xvii
Glossary	**xviii – xix**
Earth and beyond	**2–19**
Weather	2–3
Weather data	4–5
Seasons	6–7
Space – 1	8–9
Space – 2	10–11
Solar eclipse	12–13
Moon cycles	14–15
Milky Way model	16–17
Lunar craters	18–19
Energy and change	**20–37**
An energetic toy	20–21
Shoebox guitar	22–23
Insulated flask challenge	24–25
Skill tester game	26–27
An electric magnet	28–29
Moving by force	30–31
Balloon rocket	32–33
Up periscope!	34–35
A shadowy theatre	36–37
Life and living	**38–55**
Micro-organisms in food production	38–39
Balloon pump	40–41
Masters of microbes	42–43
Tracing a food chain	44–45
Gardening gurus	46–47
The year of a fruit tree	48–49
Healthy options	50–51
Upside down views	52–53
Bony hands	54–55
Natural and processed materials	**56–73**
The money bridge	56–57
Crushing columns!	58–59
Walking on eggshells	60–61
Matter mosaics	62–63
Dissolving rock	64–65
Separating colours	66–67
Making a gas	68–69
Rotten rust	70–71
Ice-cream insulator	72–73

Teacher information

Each book in the *Hands-on science* series is divided into four science topics:
- Earth and beyond
- Energy and change
- Life and living
- Natural and processed materials

Each section contains nine activities. Each activity is accompanied by a teachers page which includes information to assist the teacher with the activity.

One or more **objectives** are given for each activity page, providing the teacher with the focus of the activity and the behaviours pupils should demonstrate by completing the activity.

The required **materials** are listed clearly so the teacher is aware of what is needed to complete the activity.

Ideas listed under the **motivate** heading include suggested short activities or discussion topics designed to capture the pupils' attention and spark an interest in the lesson. By listening to pupil responses and through observation, teachers will become aware of their pupils' background knowledge.

A 'Before and after' framework, located on page xvii, can also be used to elicit pupils' prior knowledge on a topic.

The **experience** section provides easy to follow instructions for the hands-on activity. The accompanying worksheet may list the step-by-step instructions or be where the pupils record their observations and ideas after completing the task.

The **explain** section introduces the conceptual tools pupils need to interpret evidence and construct explanations, allowing them to record, discuss or present their understanding of the scientific concept experienced. It also provides the teacher with important background information about the topic (highlighted in the box).

Another opportunity for pupils to display their understanding of the concept is offered in the **apply** section. It allows pupils to apply their new knowledge and understanding to a different situation.

The **review and reflect** section asks pupils to complete an activity that evaluates their understanding of the concept. The teacher can use the result as evidence for assessment, demonstrating if understanding of learning objective has been achieved.

Answers may be included, where appropriate.

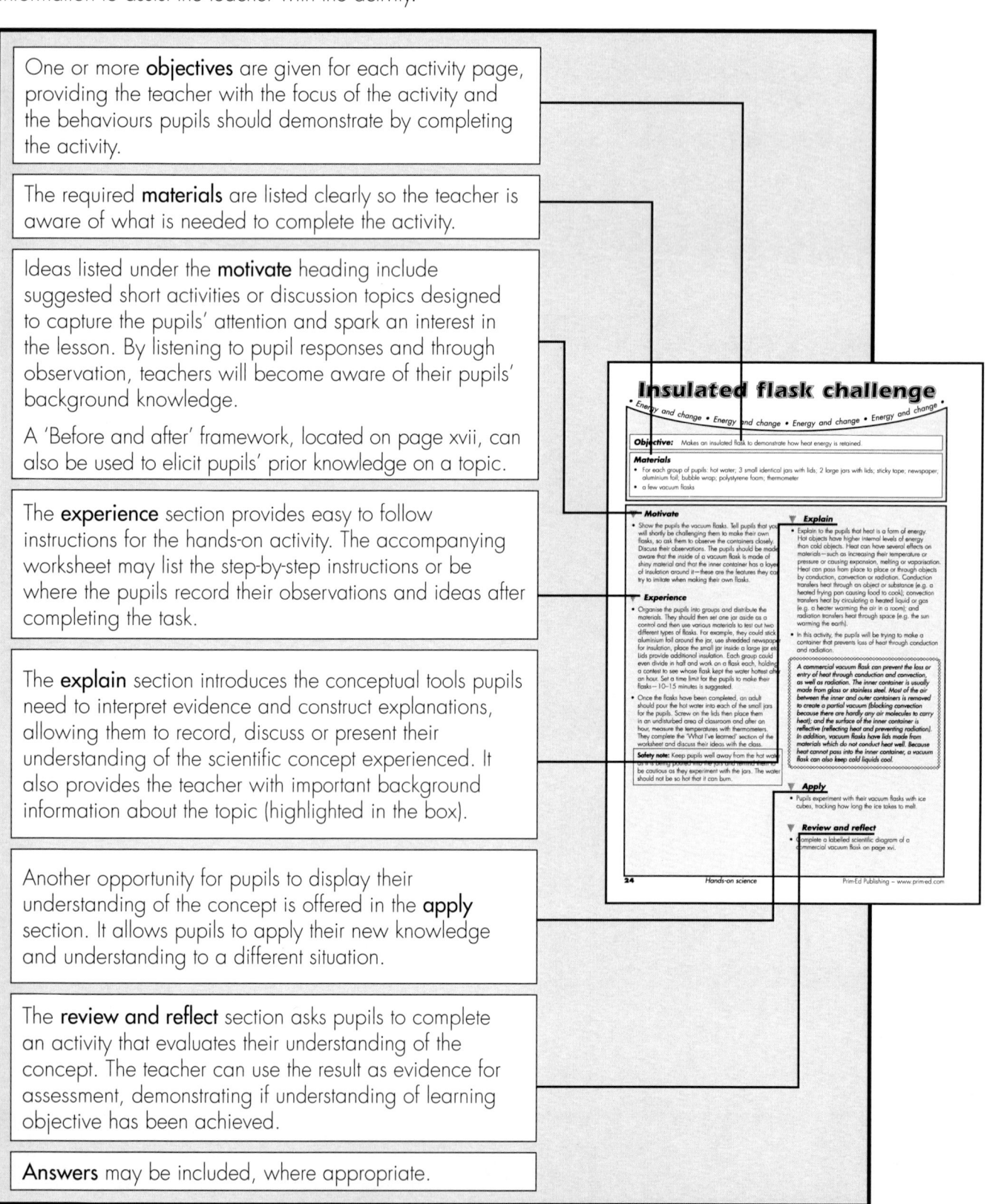

Teacher information

The pupil pages contain a variety of activities. They may be the focus of the lesson, providing step-by-step instructions to complete the hands-on experience, or provide a structure for the pupils to record their observations, investigations, results and discoveries.

The **task** is clearly stated at the top of the page, providing a focus for the pupils.

If an activity requires pupils to use material or a tool that is a **possible safety hazard** (such as hot water), the worksheet reminds them to be cautious.

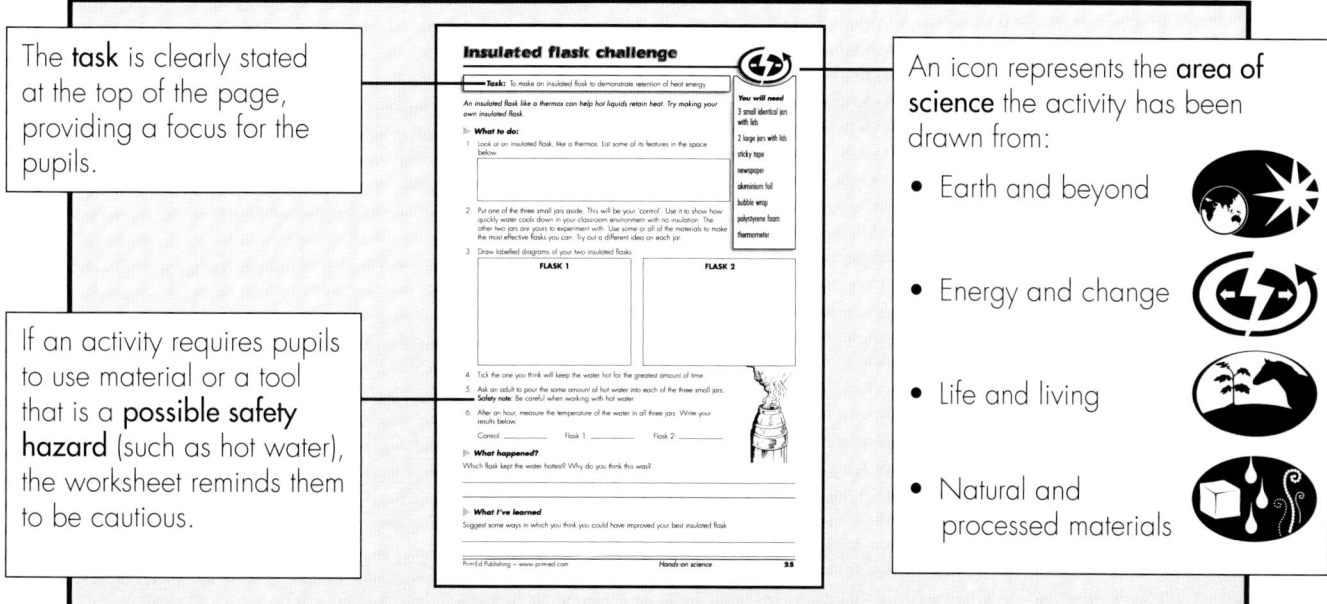

An icon represents the **area of science** the activity has been drawn from:

- Earth and beyond
- Energy and change
- Life and living
- Natural and processed materials

Frameworks

Located at the front of the book are frameworks pupils can use to display their understanding of a scientific concept and experience, and to reinforce concepts learned. The frameworks can be used to plan a new investigation or to reflect upon a completed activity. As pupils write science reports, recounts and investigations, they will be integrating science with literacy. References to relevant frameworks can be found on the accompanying teachers page of a hands-on activity.

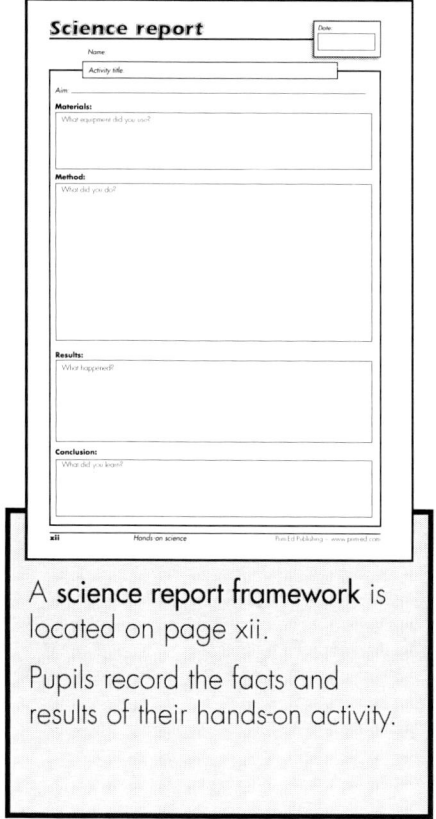

A **science report framework** is located on page xii.

Pupils record the facts and results of their hands-on activity.

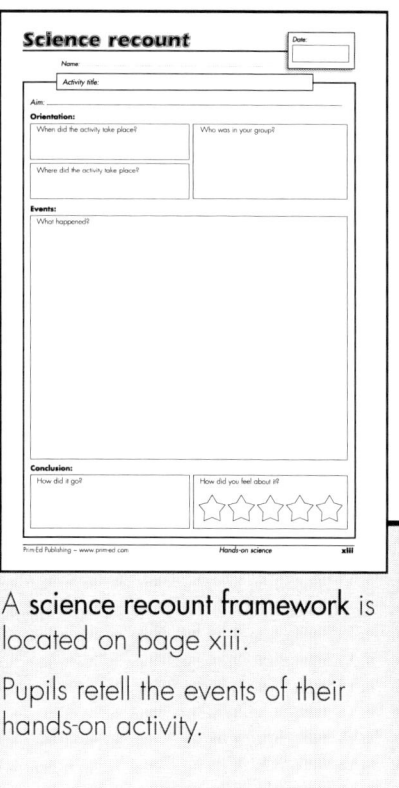

A **science recount framework** is located on page xiii.

Pupils retell the events of their hands-on activity.

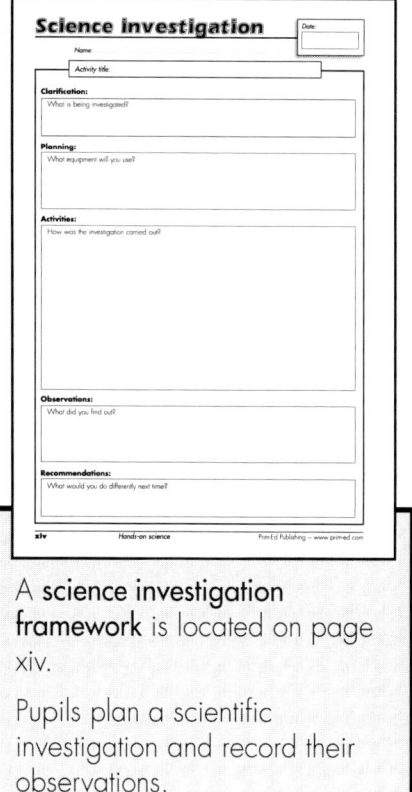

A **science investigation framework** is located on page xiv.

Pupils plan a scientific investigation and record their observations.

Teacher information

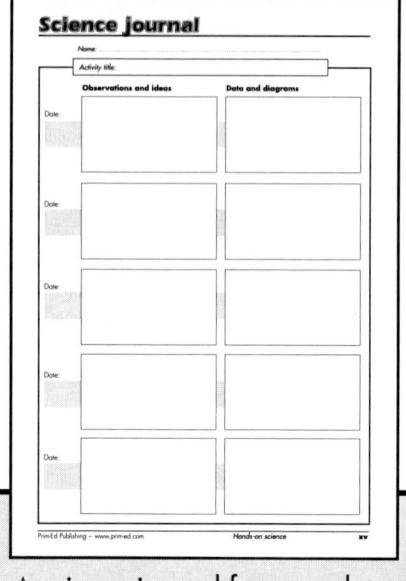

A **science journal framework** is located on page xv.

Pupils keep a dated record of observations and reflections.

A **scientific diagram framework** is located on page xvi.

After completing a hands-on activity, pupils draw a diagram.

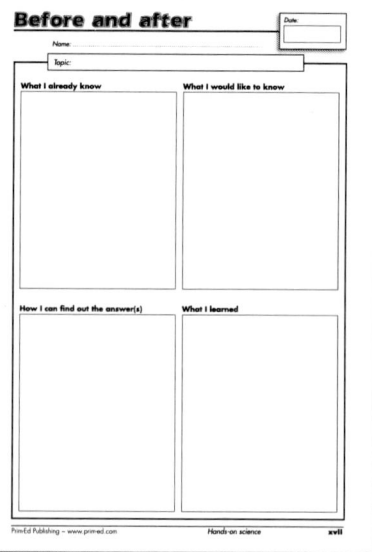

A **before and after chart** is located on page xvii.

Pupils record their prior knowledge of a topic and write questions they hope to answer. Pupils then reflect upon this after completion.

Why 'hands on'?

Hands-on learning is 'learning by doing'. A hands-on approach requires pupils to become active participants in their learning. Pupils investigate and test basic scientific principles by experimenting, creating, designing, cooking and much more, gaining an understanding of the concepts from their experiences.

Many believe that information gained through hands-on learning is remembered and retrieved better, allowing it to be transferred to other situations more easily.

Hands-on learning motivates pupils and engages them in their learning. They develop a curiosity and are interested to know 'why' something occurs. Instead of being told 'why', pupils see it for themselves, directly observing science in action.

Hands-on learning encourages questioning about the events pupils observe and the results they achieve. Pupils improve their scientific skills, such as measuring, observing, predicting and inferring.

Most of the hands-on activities in the book are conducted in groups. Collaborative learning encourages pupils to communicate clearly and express their ideas about science.

Safety

The activities in the *Hands-on science* series are safe for pupils. However, accidents can, and do, happen and so safety precautions for certain activities are given on the teachers page. Some activities also have a 'Safety note' written on the worksheet to remind pupils. It is imperative that the teacher is aware of possible safety precautions prior to an activity. If careful supervision is required during a lesson, it may be best to organise an additional adult to be in the classroom for that activity.

Science safety tips:

- Try the activity yourself before you present it to the class.
- Ensure that all groups understand the instructions; all pupils are organised and focused on the task.
- Make sure pupils are within view at all times.
- Do not hand out equipment until it is required.
- Remind pupils that they should never taste or smell any materials in a science experiment unless permission is granted by the teacher.
- If an activity is conducted in an outside area, visit the site before hand to ensure it is safe and that examples of what is to be observed are present.

Teacher information

Assessment

An assessment objective for each activity is located on the teachers page. It can be transferred to the assessment proforma on page x.

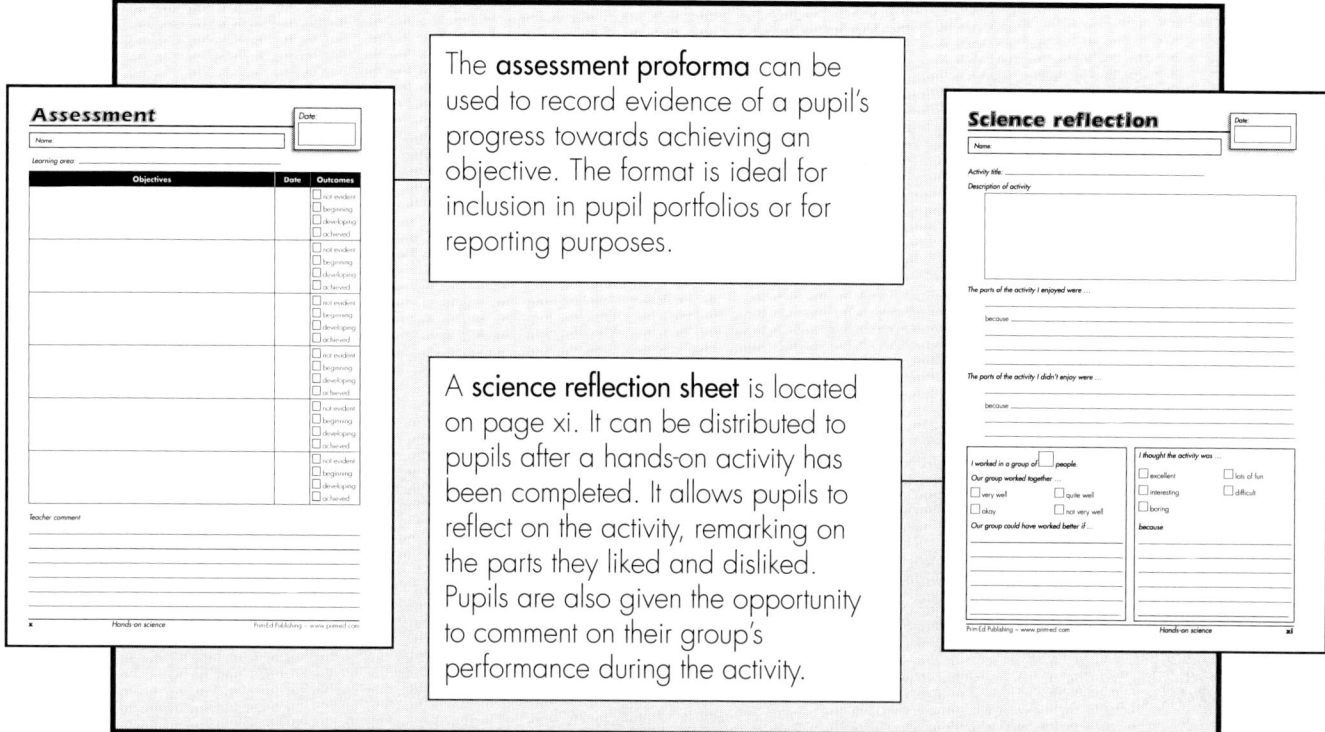

The **assessment proforma** can be used to record evidence of a pupil's progress towards achieving an objective. The format is ideal for inclusion in pupil portfolios or for reporting purposes.

A **science reflection sheet** is located on page xi. It can be distributed to pupils after a hands-on activity has been completed. It allows pupils to reflect on the activity, remarking on the parts they liked and disliked. Pupils are also given the opportunity to comment on their group's performance during the activity.

Curriculum links

	Objectives	Pages
Northern Ireland – The World Around Us – KS2	• know our place in the universe	8–19
	• investigate use of lens and design and make model telescopes	14–15 & 52–53
	• learn about a range of energy sources at home and at school	20–21
	• investigate materials as to whether they are insulators or conductors	24–25 & 72–73
	• learn about sources of energy in a variety of machines	26–27
	• construct simple circuits using components such as switches and bulbs and batteries	26–27
	• investigate how forces can affect the movement and shapes of objects	30–33
	• investigate the reflection of light from mirrors and other shiny surfaces	34–35
	• find out that when light travelling from a source does not pass through materials, shadows are formed	36–37
	• learn how to order living things in a simple food chain	44–45
	• investigate a local habitat and the relationships between plants and animals there	44–47
	• investigate conditions necessary for growth	46–49
	• learn about the life cycle of a flowering plant and how pollen is transferred	48–49
	• learn about the factors that contribute to good health	50–51
	• understand that humans have skeletons and muscles to support their bodies	54–55
	• investigate the properties of materials and how they relate to their uses	56–63
	• know that when materials are changed it may be a desirable or undesirable change	64–65
	• investigate the changes of state brought about by heating and cooling, and making a gas	68–69
	• investigate how rusting can be controlled	70–71
	• investigate materials as to whether they are insulators or conductors	72–73

Curriculum links

Republic of Ireland – Science – 5th–6th Class

Objectives	Pages
• learn that light is a form of energy	6–7
• come to appreciate that gravity is a force	8–19
• know that light travels from a source	12–13
• explore how objects may be magnified using simple lens or magnifier	14–15
• investigate use of lens and design and make model telescopes	14–15 & 52–53
• learn about electrical energy	20–21
• experiment to establish which materials are good conductors of heat or good insulators	24–25 & 72–73
• investigate current electricity by constructing simple circuits	26–27
• investigate how magnets may be made	28–29
• recognise that a gas, such as air, occupies space, can have mass and exert pressure	32–33
• investigate how mirrors and other shiny surfaces are good reflectors of light and design and make model periscopes	34–35
• become familiar with the characteristics of fungi and bacteria	38–43
• identify the interrelationships and interdependence between plants and animals in local and other habitats	44–49
• develop concept of food chains and food webs	44–45
• become aware of the sun as a source of energy for plants through photosynthesis	46–49
• develop a simple understanding of food and nutrition	50–51
• design and make a bridge that takes account of flexibility, form, stability and strength	56–59
• become aware that air is composed of different gases	68–69
• become aware of some of the practical applications of these gases in everyday life e.g. use of carbon dioxide in fizzy drinks	68–69
• investigate the effects of light, air and water on materials, such as the rusting of iron and steel	70–71
• investigate how rusting can be controlled	70–71

Wales – Science – KS2

Objectives	Pages
• know that the sun, earth and moon are approximately spherical	6–19
• know the relative positions of the sun, earth and moon in the solar system	10–11
• recognise how the position of the sun appears to change during the day and how shadows change as this happens	12–13
• know that the earth orbits the sun once a year and that the moon takes approx 28 days to orbit the sun	14–15
• know that sound is made when objects e.g. strings on an instrument, vibrate	22–23
• know that temperature is a measure of how hot or cold something is	24–25
• recognise that some materials are better thermal insulators/conductors than others	24–25 & 72–73
• know that a complete conducting circuit, including a battery or power supply, is needed for a current to flow to make electrical devices work	26–27
• know that forces can make things speed up, slow down or change direction	30–31
• know that most of the light falling on shiny surfaces and mirrors is reflected	34–35
• know that light cannot pass through some materials and this forms shadows	36–37
• know that food chains show feeding relationships in an ecosystem and that nearly all food chains start with a green plant	44–45
• recognise the process of pollination and the transport of pollen	46–49
• know that an adequate and varied diet is needed to keep healthy	50–51
• know that humans have skeletons and muscles to support and protect their bodies	54–55
• compare everyday materials on the basis of their properties and to relate these properties to everyday materials	56–63
• recognise the differences between solids, liquids and gases in terms of their properties	62–63
• explore changes in materials and recognise those that can be reversed and those that cannot	64–71
• know that heating or cooling materials can cause them to change	72–73

Curriculum links

England – Science – KS2

Objectives	Pages
• know that the sun, earth and moon are approximately spherical	6–19
• recognise how the position of the sun appears to change during the day and how shadows change as this happens	12–13
• know that the Earth orbits the Sun once each year and that the Moon takes approx 28 days to orbit the sun	14–15
• know that sounds are made when objects vibrate	22–23
• know that temperature is a measure of how hot or cold things are	24–25
• recognise that some materials are better thermal insulators than others	24–25 & 72–73
• construct circuits, incorporating a battery, to make electrical devices work	26–27
• know that objects are pulled downwards because of the gravitational attraction between them and the Earth	30–31
• know about friction as a force that slows moving objects	30–31
• know that light is reflected from surfaces	34–35
• know that light cannot pass through some materials and this leads to the formation of shadows	36–37
• know that micro-organisms are living organisms that are often too small to be seen, and that they may be beneficial or harmful	38–43
• use food chains to show feeding relationships in a habitat and know that nearly all food chains start with a green plant	44–45
• know how plants are suited to their environment	46–47
• know about the parts of the flower and their role in the life cycle	48–49
• know about the importance of an adequate and varied diet for health	50–51
• know that humans have skeletons and muscles to support their bodies	54–55
• compare everyday materials on the basis of their properties including strength and to relate these properties to everyday materials	56–61
• recognise the differences between solids, liquids and gases in terms of ease of flow and maintenance of shape and volume	62–63
• dissolve solids and describe changes when materials are mixed	64–65
• know that non-reversible changes result in the formation of new materials	68–71
• describe changes that occur when materials are heated or cooled	72–73

Scotland – Science – Second (P6/P7)

Objectives	Pages
• use simple models to communicate understanding of size, scale and relative motion in our solar system	6–7 & 10–11
• research features of space and describe them to others	8–9
• demonstrate and describe energy transfers in everyday situations and devices	20–21
• construct a simple musical instrument to demonstrate that sounds are produced by vibrations	22–23
• compare the thermal insulating properties of materials and choose the most appropriate material for a particular purpose	24–25 & 72–73
• build a game that uses simple components in a series circuit and explain in simple terms why the circuit works	26–27
• investigate how friction affects movement	30–33
• investigate the properties of light	34–37
• know how microscopic living things can be used to produce and break down foods	38–43
• appreciate the contribution made by individuals to scientific discovery and the impact this has made on society	42–43
• use knowledge of food chains and food webs	44–45
• explain plant reproduction	48–49
• make informed decisions to allow me to maintain a healthy lifestyle	50–51
• research the structure and function of the eyes	52–53
• evaluate the effectiveness of a material for its purpose	56–59
• make and test predictions about solids dissolving in water	64–65
• complete activities which demonstrate simple chemical reactions safely using everyday 'kitchen chemicals'.	69–71

Assessment

Date:

Name:

Learning area: _____

Objectives	Date	Outcomes
		☐ not evident ☐ beginning ☐ developing ☐ achieved
		☐ not evident ☐ beginning ☐ developing ☐ achieved
		☐ not evident ☐ beginning ☐ developing ☐ achieved
		☐ not evident ☐ beginning ☐ developing ☐ achieved
		☐ not evident ☐ beginning ☐ developing ☐ achieved
		☐ not evident ☐ beginning ☐ developing ☐ achieved

Teacher comment

_____
_____
_____
_____
_____

Science reflection

Date:

Name:

Activity title: _____

Description of activity

The parts of the activity I enjoyed were …

because _____

The parts of the activity I didn't enjoy were …

because _____

I worked in a group of ☐ people.

Our group worked together …

☐ very well ☐ quite well
☐ okay ☐ not very well

Our group could have worked better if …

I thought the activity was …

☐ excellent ☐ lots of fun
☐ interesting ☐ difficult
☐ boring

because

Prim-Ed Publishing ~ www.prim-ed.com	Hands-on science

Science report

Date:

Name: ....

Activity title:

Aim: _____

Materials:

What equipment did you use?

Method:

What did you do?

Results:

What happened?

Conclusion:

What did you learn?

Science recount

Date:

Name: ....

Activity title:

Aim: _____

Orientation:

When did the activity take place?

Who was in your group?

Where did the activity take place?

Events:

What happened?

Conclusion:

How did it go?

How did you feel about it?

☆ ☆ ☆ ☆ ☆

Science investigation

Date:

Name: ....

Activity title:

Clarification:

What is being investigated?

Planning:

What equipment will you use?

Activities:

How was the investigation carried out?

Observations:

What did you find out?

Recommendations:

What would you do differently next time?

Science journal

Name: ....

Activity title:

Observations and ideas **Data and diagrams**

Date:

Date:

Date:

Date:

Date:

Prim-Ed Publishing ~ www.prim-ed.com   Hands-on science   **xv**

Scientific diagram

Date:

Name: ....

Use this checklist to help you draw your scientific diagram.

- ☐ Title
- ☐ Sharp lead pencil
- ☐ Ruler used for all straight lines
- ☐ Diagram is large enough
- ☐ All parts are labelled
- ☐ Labels spelt correctly

How did you do? ☆ ☆ ☆ ☆ ☆

Before and after

Name: ....

Date:

Topic:

What I already know

What I would like to know

How I can find out the answer(s)

What I learned

Glossary

acid	a chemical compound containing hydrogen which can neutralise alkalis.
acidity	the amount of acid present in a substance.
alkali	a chemical compound containing hydroxide which can neutralise acids.
antibiotic	a drug used to kill bacteria.
antiseptic	a substance used to kill germs.
axis	a straight line running through an object; the Earth's axis is an imaginary line that runs through the North and South Poles.
bacteria	microscopic organisms.
battery	a source that stores electrical energy.
blossom	the flower of a fruit tree.
bud	undeveloped flower or leaf before fully opened.
bulb	glass container with incandescent filament that glows when electrical current passes through it.
buzzer	signalling device which produces sound by the vibration of moving parts when a magnetic field is created by electrical current.
carbon	a naturally occurring element.
carbonate	substance that reacts with an acid to form carbon dioxide.
carpals	small bones located in the wrist.
change (of state)	to change a substance from a solid to a liquid to a gas and back again.
chemical change	irreversible change caused by alteration to the chemical structure of a material.
chromatography	a separation technique where substances are dissolved in a solvent and dispersed at different rates along special paper.
conduction	the transfer of heat or electricity through a material.
consumer	those in a food chain who eat others.
control	a variable that is held constant in an experiment.
convection	the transfer of heat through movement of heated liquid or gas.
cooling	the effect of reducing temperature.
corrode	the eating away of the surface of a metal by a chemical reaction. (rust)
crater	rounded hollow in the Earth, Moon etc. formed from the impact of a meteorite.
crystals	solid particles remaining after evaporation of water from a solution.
diameter	the length of a straight line that is completed at both ends by the circumference and passes through the centre of the circle.
dissolve	to mix a soluble solute in a solvent.
electrical circuit	a closed route carrying an electric current.
electrical conductivity	the property of allowing electricity to pass through a substance.
electrical conductor	a material that electricity can flow through.
electrical current	the flow of electrical charges.
electromagnet	a temporary magnet created when an electric current flows through an iron core.
evaporation	the changing of a liquid into its gaseous state by way of heat.
fair test	a science experiment in which all the variables need to stay the same except for the one being tested.
fermentation	the action of yeast on sugar to produce carbon dioxide gas.
fertilisation	combining of male and female cells to create new life.
food chain	the sequence of living things dependent on one another for food.
friction	the force that acts against a moving object.
galaxy	a large collection of stars, dust and gas.
gas	a state of matter in which molecules are free to fill the whole space of a given container.
germinate	to sprout and send out shoots.
gravity	a force of attraction between a very large object and smaller objects.
habitat	where an organism lives.
heartbeat	the sound the heart makes as it pumps blood around the body.
heating	the effect of increasing temperature.

Glossary

image	picture of an object as light rays pass through a lens or are reflected by a mirror.
immunity	protection from disease.
infection	an attack on an organism by germs or disease.
insulator	material that prevents the flow of electricity through it.
irreversible change	cannot be returned to its former state.
life cycle	the development of life from fertilisation to the production of a new generation by a species.
light-year	the distance light can travel in one year; a unit of measurement commonly used by astronomers.
liquid	state of matter in which molecules are free to move among themselves, but not into air.
lunar	having to do with the Moon.
material	the substance something is made from.
melt	when a substance is changing from a solid to a liquid state.
metacarpals	long bones in the palm of the hand.
meteorite	small body of stone or metal interplanetary matter (usually from a comet), that travels through space and impacts the Earth's (or Moon's) surface.
microbiology	study of microscopic organisms.
micro-organism	a tiny plant or animal organism, invisible to the naked eye.
microscope	viewing apparatus with powerful magnifying lenses capable of showing micro-organisms in the presence of light.
Milky Way	a spiral galaxy in which the Earth and its sun can be found.
molecule	a very small particle of a substance.
mould	micro-organism that in large numbers produces a furry covering on animal or vegetable matter.
object	the actual thing from which light rays bounce to allow it to be seen.
organism	a living thing, animal or plant.
particles	very small parts of a material/compound.
pH	degree of acidity.
phalanges	the bones of the fingers and toes.
photosynthesis	process by which plants use sunlight, water and carbon dioxide to make their food.
physical change	reversible change to a substance from one state to another.
pitch	how low or high a sound is; musical notes with a low pitch have a slower frequency of vibration than ones with a higher pitch.
pollen	male cells of a flower.
pollination	transfer of pollen from one plant to another to enable fertilisation.
producer	plants that provide food through photosynthesis for the next level of the food chain.
radiation	the transfer of heat through space.
rust	the red/orange coating that forms on the surface of iron when exposed to air and moisture.
saturation	state in which no more solute can be dissolved in a solvent.
seed dispersal	the distribution of seeds by wind, insects and animals to enable growth in other locations.
separation	two or more materials physically removed from one another.
solid	state of matter in which molecules are rigidly held together.
solubility	the extent to which a solute can dissolve in a solvent.
solute	the material being dissolved in a solvent.
solution	a substance in which a solute is dissolved in a solvent.
solvent	a material which dissolves a solute to form a solution.
soundwaves	vibrations of particles in a solid, liquid or gas by which sounds are transmitted.
speed	rapidity of movement.
thrust	a reactive force that pushes a rocket or jet aeroplane forward.
vaccine	a substance made from disease causing bacteria which, when injected, gives immunity from that disease.
weather	the day-to-day changes and effects of wind, temperature, clouds, moisture and air pressure.
weathering	a physical or chemical process that breaks down or wears away rock.
yeast	micro-organisms which produce carbon dioxide gas through the process of fermentation.

Weather

Earth and beyond • Earth and beyond • Earth and beyond • Earth and beyond

Objective: Develops an understanding of weather maps.

Materials

- weather maps collected out of newspapers, collection of weather reports from TV, recorder, blank tapes and/or CDs

▼ Motivate

- Display to pupils several weather maps and collection of television weather reports.
- Discuss with pupils the features shown on the weather maps. Explain what the different words/symbols represent.
- Discuss the manner and tone of weather reporters. Point out to pupils the clear and concise way reporters conduct the weather report.

▼ Experience

- Organise pupils into pairs.
- Read with pupils the explanatory text in the box below, study the map and the key explaining the symbols on the map.
- Pairs work together to write and record a national weather summary for radio.

▼ Explain

- Discuss how a weather map works. In particular, talk about warm and cold fronts, low- and high-pressure systems and isobars.

A cold front is the area where a cold air mass replaces a warm air mass. The air behind a cold front is colder and drier than the air ahead of it. A cold front is represented on a weather map by a solid line with triangles pointing towards the warmer air and the direction it is moving. On coloured maps, cold fronts are usually represented by a solid blue line.

A warm front is the area where a warm air mass replaces a cool air mass. The air behind a warm front is warmer and more moist than the air in front of it. A warm front is represented by a solid line with semicircles pointing towards the colder air and the direction it is moving. On coloured maps, warm fronts are usually represented by a solid red line.

An isobar is a line connecting locations of equal barometric pressure. Isobars are drawn at intervals of 2 or 4 millibars and can be used to identify 'highs' and 'lows'. The pressure in a 'high' is greater than the surrounding air. The pressure in a 'low' is lower than the surrounding air. The strength of wind is determined by changes in pressure (the pressure gradient). A large change in pressure over a short distance indicates strong winds—displayed by isobars drawn close together.

▼ Apply

- Ask the class to look out the window and observe the weather.
- Pupils create a weather report for television, taking care to use correct terminology. Pupils record report with a partner using a video camera. Replay it for the class to review.

▼ Review and reflect

- Complete a science recount, found on page xiii.

▼ Answers

While answers will vary, each should include:

Bowden: S-SE winds, varying, then changing to S-SW; cool to cold temperatures becoming warmer; drizzle or light/moderate rain gradually easing/continuing.

Chatsfield: S-SW winds, becoming gusty then changing to W-NW; warm followed by a sudden drop in temperature then becoming steadily cooler; brief showers followed by heavy rain and the possibility of hail, lightning and thunder, before easing to showers and then clearing.

Weather

Task: To record a national weather bulletin suitable for radio.

Weather charts are made up of curved lines which are drawn on a geographical map. These are used to show the weather features of a particular geographical area. Some features shown on these charts include:

- atmospheric pressure; which consist of isobars (lines of equal pressure) drawn around lows (L) and highs (H). (When isobars are close together, winds are stronger.)
- fronts and troughs; which highlight the areas of most significant interest.

You will need

recorder

blank tape or CD

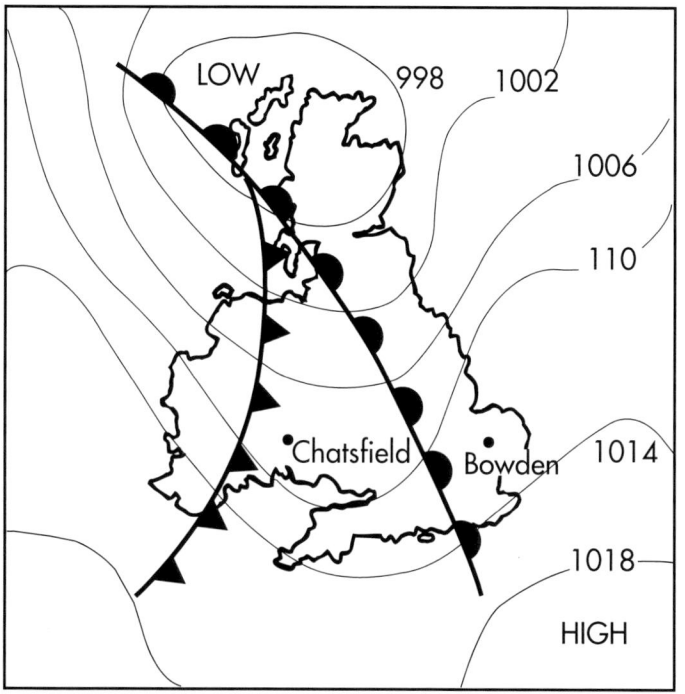

	Before	During	After
Warm front			
winds	S–SE	variable	S–SW
temperature	cool/cold, warming	steady rise	warmer
precipitation	light/moderate rain, drizzle	drizzle or none	none or light rain or showers
Cold front			
winds	S–SW	gusty	W–NW
temperature	warm	sudden drop	dropping steadily
precipitation	brief showers	heavy rain, hail, thunder, lightning	showers then clearing

Key

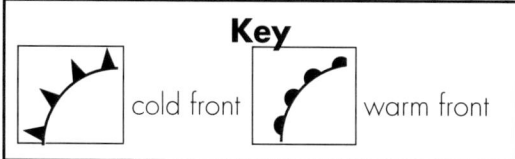

cold front — warm front

What to do:

1. Look at the weather map and key above.
2. Write a national weather forecast for radio using the information from the map.

Introduction (warm and friendly):
Weather conditions (formal and concise):
Bowden
Chatsfield
Conclusion (a general comment about the day ahead):

3. Record your weather forecast onto a tape or CD and play for the class.

Weather data

Earth and beyond • Earth and beyond • Earth and beyond • Earth and beyond

Objective: Records and analyses weather data in order to make a prediction.

Materials
- news report or newspaper showing the predicted weather for the week, thermometer and rain gauge for each group

▼ Motivate
- Show pupils various news reports or newspaper articles predicting the weather for the week. Look at the symbols used.
- Discuss with pupils how the predicted weekly reports often change each day. Talk about why things change. Weather is reasonably changeable and therefore unpredictable.

▼ Experience
- Organise pupils into groups.
- Ensure the pupils understand the key explaining the symbols used when recording weather information.
- Each group decides the best place to locate their thermometer and rain gauge for the week. Ensure they are easily accessible but not in the way of regular play or sport; i.e. It wouldn't be a good idea to locate the instruments in the middle of the school playground.
- Decide on the time of the school day the data will collected and recorded. Stick to this routine as closely as possible.
- Pupils collect data each day and record the information in the table.
- From the data collected, pupils record their predictions for the weekend. This could be produced as a computer presentation in the form of a weather report.
- Pupils rate their predictions after the weekend. Discuss accuracy or inaccuracy of predictions. (Data needs to be collected at the same time of day.)

▼ Explain
- Discuss weather data recording and analysis with pupils.

> Weather data is collected all around the world from a variety of sources (weather stations, weather balloons and satellites). Predictions are made based on data collected rather than speculation.
>
> Weather conditions on Earth rely on many elements: wind, temperature, moisture, air pressure and humidity. People rely on accurate weather predictions for many different reasons: Knowing about the weather (especially wind) is important for pilots and the safety of their passengers. Anglers need to know whether to venture out to sea because storms could mean conditions are unsafe. A farmer's livelihood depends greatly on the weather and climate patterns.

▼ Apply
- Establish a pupil weather watching team responsible for reporting local weather conditions to the school. This could allow teachers and pupils to plan a day's events according to predicted weather conditions.

▼ Review and reflect
- Pupils complete a science reflection sheet on page xi.

Weather data

Task: To record and analyse local weather for one week in order to predict the weekend weather.

You will need

thermometer and rain gauge for each group

What to do:

Complete the weather chart below by recording data at the same time each day for one week. Use these keys to help you.

Cloud cover
- ○ no clouds
- ◔ 1/8
- ◔ 2/8
- ◑ 3/8
- ◐ 4/8
- ◐ 5/8
- ◕ 6/8
- ◕ 7/8
- ● 8/8 completely overcast

Wind
- 0 km/h (smoke rises)
- 5 km/h (smoke drifts)
- 10 km/h (flags stir)
- 20 km/h (leaves move)
- 40 km/h (trees sway)
- 50 km/h (flags beat)
- 60 km/h (flags extend)
- 75 km/h (difficult to walk)
- 115 km/h (trees uprooted)

Weather chart

	Maximum temp. (°C)	Rainfall (mm)	Cloud cover	Wind speed
Monday				
Tuesday				
Wednesday				
Thursday				
Friday				

What will happen?

Use your observations from over the week to predict the weekend weather.

	Maximum temp. (°C)	Rainfall (mm)	Cloud cover	Wind speed
Saturday				
Sunday				

What happened?

Rate your weather predicting skills.

poor					excellent

Hands-on science

Seasons

Earth and beyond • Earth and beyond • Earth and beyond • Earth and beyond

Objective: Identifies that seasons change due to the tilt of the Earth's axis.

Materials
- photographs, pictures or video of different seasons, 1 torch and 1 piece of blank paper per group of two or three pupils, globe of the Earth

▼ Motivate
- Display to pupils various pictures or footage of the different seasons.
- Discuss with pupils the general weather conditions associated with each season. What do pupils notice about the temperature, daylight hours etc.

▼ Experience
- Organise pupils into pairs or small groups.
- Distribute one torch and one piece of blank paper to each group.
- Direct pupils to Step 1, allowing groups time to record their observations. Repeat with Step 2.
- Demonstrate the same idea using the globe and the torch. Allow pupils time to discuss this at Step 3 and record their thoughts.

▼ Explain
- Discuss the relationship between the Earth's seasons and tilt of the Earth's axis with pupils.
- Pupils complete the 'What I've learned' questions.

The seasons on Earth are governed by the tilt of the Earth's axis as it travels in space around the sun. When the Earth is tilted so that the sun is directly over the Tropic of Cancer (near 21 June), less sunlight gets scattered before reaching the ground because it has less distance to travel through the atmosphere and the high angle of the sun produces longer days. This is when the Northern Hemisphere has its summer. The opposite is true in the Southern Hemisphere, where the low angle of the sun produces shorter days and large amounts of energy are dispersed by the atmosphere. This is the date of the Southern Hemisphere's winter solstice because it is tilted furthest away from the sun. Near 21 December, when the sun is positioned directly over the Tropic of Capricorn, the Southern Hemisphere is closest to the sun, with little scattering of the sun's rays and the sun's high angle producing longer days. The Northern Hemisphere is tilted away from the sun, producing short days and a low sun angle.

During the equinoxes (about 21 September and 21 March), the sun is directly above the Equator and the sun's energy is balanced equally between the Northern and Southern Hemispheres. On this day, the Earth's hemispheres experience day and night of approximately the same length.

▼ Apply
- Create a model of the Earth's rotation around the sun, including the tilting of the Earth's axis, to recreate the seasons. Discuss the season in each hemisphere while recreating the scenario.

▼ Review and reflect
- Complete a 'Before and after' chart on page xvii.

▼ Answers
What to do:

Step 1: All of the torch light is contained in the illuminated circle, which gives a more direct light which means greater heat and hotter temperatures.

Step 2: The light is spread out over a greater distance which means the light is less direct and gives less heat and cooler temperatures.

Step 3: Teacher check

What I've learned

1. Summer; The heat and light is more concentrated and, therefore, hotter.
2. The sunshine is spread out creating different degrees of light and heat over the Earth's surface.

Seasons

Task: To observe how the Earth's tilt governs the seasons.

You will need
torch
blank piece of paper

▶ **What to do:**

1. Shine the torch directly at the piece of paper.

   What do you see?

2. Slowly tilt the paper.

   What do you see?

3. Consider that the torch represents the sun and the paper represents the Earth. In your group, discuss how you think this tilting affects the seasons on Earth and record your thoughts.

   My thoughts

Repeat the process if you need to when answering these questions.

▶ **What I've learned**

1. Which of Earth's seasons do you think was represented by Step 1 of the activity?

   ☐ summer ☐ autumn ☐ winter ☐ spring

   Explain why _____

2. In what ways would Earth be affected in Step 2 of the activity?

   _____

Space – 1

Earth and beyond • Earth and beyond • Earth and beyond • Earth and beyond

Objectives:
- Applies research skills to develop a greater understanding about a given planet or star in the solar system.
- Works cooperatively in a group to display information in an appropriate style.

Materials
- photographs of space, video footage of space (a search of the Internet can turn up some great websites about space), display card; paper or access to a computer that allows computer presentations; video or DVD recorder

▼ Motivate
- Display to pupils various pictures or footage of space.
- Discuss and list what the pupils already know about the solar system.
- Complete a 'Before and after' chart (page xvii) about the solar system.

▼ Experience
- Organise pupils into nine groups. Each group is assigned one of the eight planets and the final group is given the sun.
- Read through the worksheet with pupils and explain in detail what each group is going to do.
- Pupils work together to research and record information in note form about their planet or star. Note: A number of the categories do not apply to the sun.
- Once all information is gathered, pupils decide how they would like to present the information to the rest of the class. Some groups may choose to make a booklet or a poster, while others may choose to make a computer presentation or a short documentary.
- Pupils then present their information to the class for evaluation. Other pupils should assess the presentation, based on clarity and accuracy of information provided.

▼ Explain
- Discuss the solar system with pupils.

> *The solar system is the name given to the area of space that contains our star and its eight planets. The star is commonly called the sun and is the centre of the system. It is over seven hundred times the size of all of the system's planets and moons combined. This enormous mass creates a gravitational pull that keeps the planets in position while circling the sun. This is the same force that keeps the moon orbiting the Earth.*
>
> *The planets in order from the sun are: Mercury, Venus, Earth, Mars, Jupiter, Saturn, Uranus and Neptune.*
>
> *Other than the planets and the sun, the solar system also contains moons, asteroids, meteoroids, comets and dust.*

▼ Apply
- Create a television commercial designed to attract visitors to your planet or star. Present this to the class.

▼ Review and reflect
- Complete a science recount chart on page xiii.

Space – 1

Task: To work as a group to research a part of the solar system and present findings to the class.

You will need

access to the library or the Internet

display materials or resources

▶ What to do:

1. Working as a group, research the following relevant information about your planet or star.
2. Record your notes below before presenting your findings to the class in a display style of your choice.

Name _____

Physical description

Surface: _____
_____
_____

Atmosphere: _____
_____
_____

Moons:

Rings:

Special features: _____
_____

Facts and figures

Size: _____   Mass: _____

Distance from sun: _____   Surface temperature: _____

Length of day: _____   Length of year: _____

Gravity: _____   Orbital direction: _____

Related Roman god/goddess _____
_____
_____

Astrology links _____
_____
_____

Prim-Ed Publishing ~ www.prim-ed.com   Hands-on science

Space – 2

Earth and beyond • Earth and beyond • Earth and beyond • Earth and beyond

Objective: Understands the relationship between the sun and the planets in the solar system.

Materials

- photographs of space, video footage of space, (a search of the Internet can turn up some great websites about space); headbands; appropriate gear to make labels, headpieces or costumes to represent the different planets

▼ Motivate

- Provide pupils with various pictures or footage of space.
- Discuss and list what the pupils already know about the solar system.
- Complete a 'Before and after' chart (page xvii) about the solar system.

▼ Experience

- Organise pupils into groups of nine.
- Read through the worksheet with pupils and explain in detail what each group is going to do.
- Pupils work together to assign planets to each member of their group.
- Allow pupils time to prepare headbands or costumes for their performance.
- Allow time for pupils to practise their performance.
- As groups are ready, they ask for a teacher evaluation. The teacher will evaluate the pupils as ready to take their performance to the class or give pointers to improve the performance.
- Groups perform their recreation for the class.

▼ Explain

- Discuss the solar system with pupils.

> The solar system contains all the planets, asteroids and comets that orbit our sun. Early astronomers used their eyes to examine the sky. Today, we use powerful telescopes to view space.
>
> Just looking at the sky, planets look like stars; however, they are much closer to us than any star. Like Earth, the other seven planets move slowly across our sky while orbiting the sun. Venus is the brightest planet and is called 'the morning star' at sunrise; at sunset, it is known as 'the evening star'. The only other planet that is visible to the human eye is Mars.
>
> The four planets that are the closest to the sun— Mercury, Venus, Earth and Mars—are made of rock and iron. Jupiter, Saturn, Uranus and Neptune are balls of gas.

▼ Apply

- Create a 3-D model of the solar system. Produce a computer presentation of your model.

▼ Review and reflect

- Complete a science reflection chart on page xi.

Space – 2

Task: To actively participate in a recreation of the solar system.

What to do:

1. Work as a group to recreate the solar system. You will need to prepare by:

   ☐ appointing a role to each member of your group and recording his/her role in the table below.

   You will need
   useful materials to make costumes or headbands

Sun		Mercury	
Venus		Earth	
Mars		Jupiter	
Saturn		Uranus	
Neptune			

   ☐ creating suitable costumes or headbands that make it easy for observers to tell which planet is being represented (think carefully about the special features of your planet).

2. Once you know who will present each planet and have the appropriate gear, it is time to practise 'being' the solar system. This will be a difficult task. Refer to the table below to help with the distance each person needs to be from the sun. (You will obviously need to scale this down!)

Planet	Distance from sun (km)	Diameter (km)	Length of day	Length of year	Special features	Orbital direction
Sun		1 392 000				
Mercury	58 000 000	4900	175.94 Earth days	87.97 Earth days	large craters	anticlockwise
Venus	108 000 000	12 100	243.16 Earth days	224.7 Earth days	clouds swirl around planet	clockwise
Earth	150 000 000	12 800	23.93 hours	365.242 days	inhabited	anticlockwise
Mars	228 000 000	6800	24.62 Earth hours	686.98 Earth days	appears red in colour	anticlockwise
Jupiter	778 000 000	140 000	9.83 Earth hours	11.86 Earth years	64 moons; thin ring	anticlockwise
Saturn	1 429 000 000	121 000	10.66 Earth hours	29.46 Earth years	50 moons; rings	anticlockwise
Uranus	2 875 000 000	51 000	17.24 Earth hours	84 Earth years	27 satellites	clockwise
Neptune	4 504 000 000	49 000	16.11 Earth hours	164.8 Earth years	13 moons	anticlockwise

3. Key points to remember when playing your part:

   ☐ the further from the sun you are, the slower you move (this determines the length of your year).

   ☐ consider the orbital direction you must travel.

   ☐ make sure you are also spinning on your axis to make your day and night.

   ☐ remember, the sun is like an anchor for the solar system, holding everything together.

4. Practise your solar system recreation as a group several times. When you are happy with the way your performance is developing, ask your teacher to evaluate your group. If your teacher is happy with your performance, you are ready to show the rest of your class. Enjoy!

Solar eclipse

Earth and beyond • Earth and beyond • Earth and beyond • Earth and beyond

Objective: • Makes a suitable model to demonstrate a solar eclipse.

Materials

- photos or footage of a solar eclipse; Internet; for each group: round objects to represent the Earth, sun and moon (balls, basketball, world globes etc.); torch; craft and household materials requested by pupils to create their models.

▼ Motivate

- Show pupils various pictures or footage of a solar eclipse.
- Search the Internet to find out the date and location of the last solar eclipse and when and where the next eclipse is due to occur.

▼ Experience

- Organise groups of three or four pupils.
- Read through the worksheet with pupils and allow groups to discuss and plan their model. Record planned materials and procedure in appropriate spaces on the table.
- Pupils gather relevant materials, test their model and record what happened.
- Pupils evaluate the process and what they discovered.

▼ Explain

- Discuss how solar eclipses occur with pupils.
- Ask pupils to, individually, explain in their own words why they think solar eclipses occur so infrequently. The activity should have given him/her some indication as to how difficult they are to achieve.

Conditions need to be precise in order for a solar eclipse to occur. When the moon passes between the Earth and sun during the new moon phase, the shadow of the moon blocks out the sun's light, resulting in a solar eclipse.

The moon crosses the Earth's orbital plane twice a year. These are known as eclipse seasons and are the only times of the year a solar eclipse can occur.

There are three different types of solar eclipses that can occur:

- total eclipse—the entire sun is blocked out by the central region (umbra) of the moon's shadow; the sky darkens as if it were night time.
- partial eclipse—the outer region of the moon's shadow (penumbra) passes in front of the sun; the sky dims. (How dark it becomes depends on the amount of shadow.)
- annular eclipse—this occurs when the moon is at a point in its orbit where it is in line with the sun, but the moon appears smaller than the sun. The light of the sun is not blocked, instead there is a ring-like sliver of light from the sun.

Safety note: It is dangerous to directly watch a solar eclipse as it will damage the eyes. Pinhole devices should be made.

▼ Apply

- Pupils research and present their findings on the effect of a solar eclipse on animals.

▼ Review and reflect

- Complete a science reflection chart on page xi.

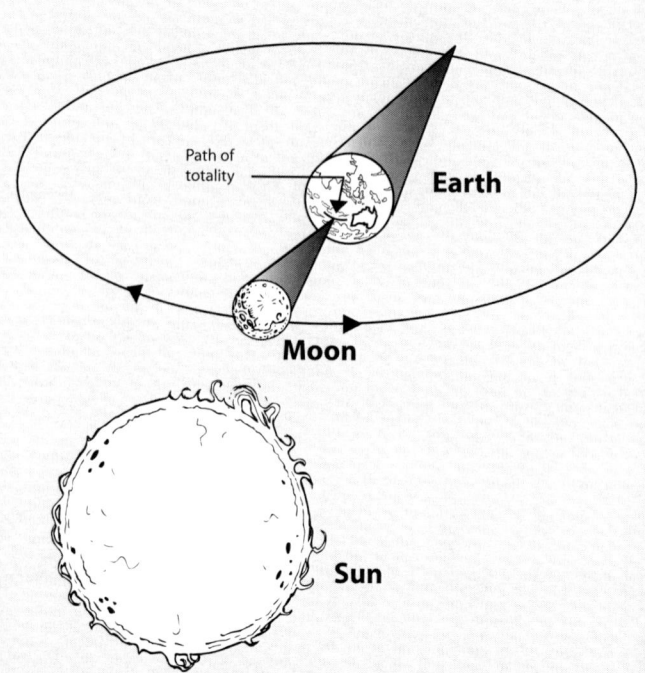

Solar eclipse

Task: To create a model that demonstrates the effect of a solar eclipse.

A solar eclipse results when the moon passes in a direct line between the Earth and the sun. The eclipse occurs because the moon's shadow (caused by the sun) travels over part of the Earth's surface and blocks out the sun's light at that point on the Earth. A solar eclipse isn't visible at every location on the Earth; a person must be in the right place at the right time (where the shadow falls) and there must be a new moon lunar phase for it to work.

You will need
torch
model to represent the Earth, sun, moon (marble, pea, basketball etc.)
craft and household materials

▶ **What to do:**
Use various materials and your knowledge of the solar system to create a model that demonstrates a solar eclipse.

People in my group:	
Materials we will use:	What we will do:

▶ **What happened?**
Draw a labelled diagram that describes the experiment.

▶ **What we learned**
After completing the task, explain how solar eclipses occur and why they are so rare.

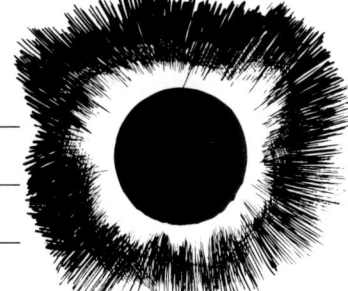

_____
_____
_____
_____

Prim-Ed Publishing ~ www.prim-ed.com   Hands-on science

Moon cycles

Earth and beyond • Earth and beyond • Earth and beyond • Earth and beyond

Objectives:
- Follows directions to construct a telescope.
- Uses the telescope to observe the lunar cycle over a period of 28 days.

Materials
- telescope, pictures/footage of the moon
- per pupil: concave lens, convex lens, 3 cardboard cylinders that slide into one another, disc of stiff cardboard to hold lens in place, glue, putty, scissors and paint

▼ Motivate
- Show pupils various pictures or footage of the moon.
- Ask pupils to list what they already know about the moon's cycle.
- Ask pupils to look at, touch, explore and use the telescope. What does it do? What do the pupils notice?

▼ Experience
- Pupils read through the worksheet and gather materials.
- Work through the steps with the pupils to make the telescope. Some pupils may need help.
- Pupils take the telescopes home and wait for the moon to appear (unless there is no moon or there is dense cloud cover). Encourage pupils to use the telescopes to become familiar with them.
- Pupils then use the telescope to observe the moon and complete the table.
- At the end of the 28-day period, organise pupils into small groups to discuss observations: What did they see? What did they notice? Was there a pattern?

▼ Explain
- Discuss the moon with pupils.

> The moon is a natural satellite of the Earth. The moon does not produce any light of its own. We can only see the moon on Earth because it reflects the sun's light. It takes approximately one month for the moon to revolve in orbit around the Earth. The moon is held in orbit by the Earth's gravitational pull.
>
> The moon is unique in the way it rotates on its axis; it keeps the same side facing Earth at all times—we never get to see the other side! The other half of the moon is always facing towards space. As the moon travels around the Earth, the amount of lit surface observable from the Earth changes. When the moon is located between the Earth and the sun, the area that is lit is facing away from the Earth. On Earth, we are unable to see the moon as the side facing us is completely in shadow. This is known as a new moon.
>
> As the moon moves in its orbit, the part of the moon's surface visible to Earth starts to receive light. We can see a small sliver of the moon (the 'C' shape that we sometimes associate with the moon). This is known as a waxing crescent.
>
> The moon continues its orbit and more of the surface of the moon that is visible to Earth receives the sun's light. First we see half of the moon (first quarter), and as it continues only a small part is in shadow (waxing gibbous).
>
> When the Earth is located between the moon and the sun, the side facing Earth is completely in the sun's light. All of the moon's surface that is visible to Earth is lit. This is known as a full moon.
>
> As the moon's travels continue, its visible surface starts to receive less sunlight. This is known as the waning part of its journey. At first, a small part of the visible side disappears into shadow (waning gibbous), then half of it (last quarter). At the last stage, only a small 'slice' of the moon can be seen (waning crescent). Finally, it starts where it began; where all of the side of the moon that we can see from Earth is in shadow and there is a new moon. Then the process or cycle starts again. This is known as the lunar cycle.

▼ Apply
- Ask pupils to research and present their findings on the effect the moon has on the Earth's tidal patterns.

▼ Review and reflect
- Complete a 'Before and after' chart on page xvii.

Moon cycles

Task: To make a telescope in order to observe the moon's lunar cycle.

▶ What to do:

1. Take the convex lens and attach it with putty to the end of the largest tube. Make sure that it is straight.

2. Take the disc of stiff card. Pierce a hole through its middle (wide enough to look through). Attach the concave lens to the card. Attach lens and eyepiece to final cardboard cylinder.

3. Slide the three cylinders into each other to make a long cylinder that can be adjusted in length.

4. Paint the outside of the telescope.

5. Look through the eyepiece and move the cylinders until you are able to see a clear image.

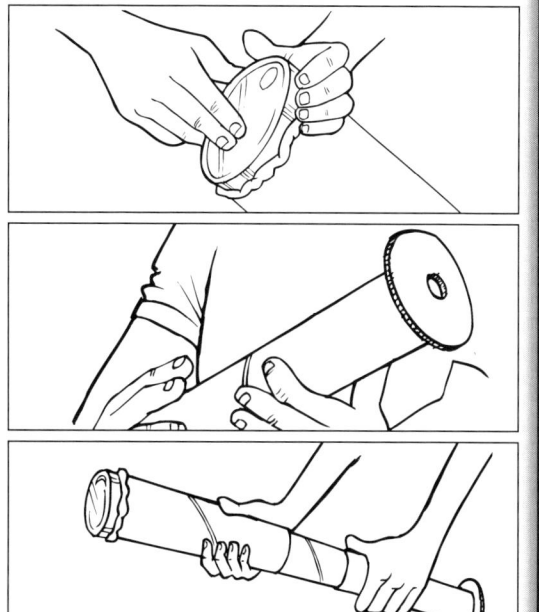

You will need

concave lens

convex lens

3 cardboard cylinders that slide into one another

disc of stiff cardboard to hold lens in place

glue

putty

scissors

paint

▶ What happened?

1. Use your telescope to view the moon each night for four weeks. Record your observations below by shading the part of the moon you are unable to see.

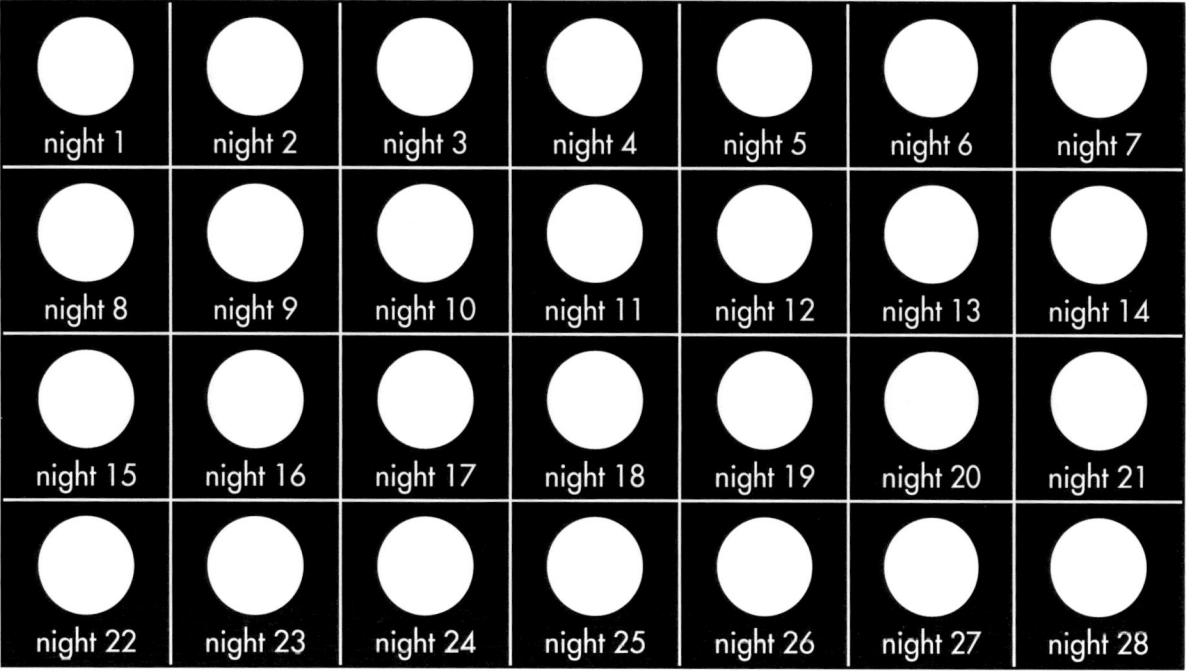

2. Discuss your observations in a small group.

Prim-Ed Publishing ~ www.prim-ed.com   *Hands-on science*   **15**

Milky Way model

Earth and beyond • Earth and beyond • Earth and beyond • Earth and beyond

Objectives:
- Makes a scale model of the Milky Way.
- Understands some of the features of the Milky Way galaxy.

Materials
- For each pupil: thick black card; scissors; cotton wool; craft glue; red, blue, silver and gold glitter; pencil; fine gold marker; ruler

▼ Motivate

- Show pupils photographs of the Milky Way and other galaxies. Ask the pupils if they know what a galaxy is (a large collection of stars, dust and gas) and the name of the galaxy our solar system is in.
- Discuss the fact that the Milky Way (and the universe) is so huge that its size is measured in light-years, rather than kilometres. A light-year is a unit of measurement commonly used by astronomers and equals the distance that light can travel in one year. One light-year is equal to approximately 9.46 trillion kilometres!

▼ Experience

- Read the information at the top of the worksheet with the class.
- Distribute the materials to each pupil. Ensure pupils read through the instructions before they begin making their models. Make sure the pupils understand that the gold dot (representing our sun) should only be about the size of a single grain of glitter.
- String could be attached to the completed models so they can be hung from the ceiling to make a display.

▼ Explain

- Once the models have been made, discuss what each part of the model represents. Some of the facts below could also be shared.

The Milky Way is a spiral galaxy that contains the Earth, our sun, clouds of dust and gas, and more than 100 billion stars. If you could view the Milky Way from above, it would look like a huge pinwheel. However, from Earth, the Milky Way looks like a white streak in the sky. Most of the stars we can see in the night sky will be in one of the 'arms' of the Milky Way.

The diameter of the Milky Way is approximately 100 000 light-years, with the central bulge being about 10 000 light-years thick. In the scale model made in this activity, each centimetre of the disc represents about 3300 light-years. Our sun is approximately 27 000 light-years from the centre of the Milky Way (about 8 centimetres from the middle of the scale model). Our solar system would be too tiny to mark on this model.

▼ Apply

- Have the pupils plan an oral presentation to explain their models of the Milky Way.

▼ Review and reflect

- Complete a 'Before and after' chart on page xvii.

▼ Answers

What I've learned

Labels should indicate the following:
- The black circle represents the disc shape of the Milky Way
- The lump of cotton wool represents the central bulge
- The teased-out pieces of cotton wool represent the spiral arms
- The gold dot represents our sun
- The blue and white glitter represents hotter stars
- The red and yellow glitter represents cooler stars.

Milky Way model

Task: To make and explain the features of a scale model of the Milky Way.

The Milky Way is a spiral galaxy that contains the Earth, our sun and more than one hundred billion stars—including all the stars we can see in the night sky. It is the shape of a thin disc, with a bulge in the middle and spiral arms of stars, dust and gas spreading out from the bulge. Generally speaking, the hotter stars in the Milky Way are found in the spiral arms and the cooler stars are found in the disc and the bulge. Our sun is a star of average temperature and is found in one of the spiral arms.

You will need

thick black card

scissors

cotton wool

craft glue

red, blue, silver and gold glitter

pencil

fine gold marker

ruler

▶ **What to do:**

Make a scale model of the Milky Way.

(a) Cut a 30 cm-diameter circle from the card. Mark the centre of the circle with a pencil.

(b) Glue a handful of cotton wool to the centre of the circle. Shape it so it is about 8 cm wide and 3 cm high. Turn the circle over and repeat on the other side.

(c) On both sides of the circle, carefully tease the cotton wool out into two or more spiral arms. Glue them onto the circle.

(d) Spread glue onto the spiral arms and sprinkle on blue and silver glitter.

(e) Spread glue onto the circle and the central lump of cotton wool and sprinkle on red and gold glitter. There should be more glitter on the cotton wool than on the circle.

(f) When the glue is dry, repeat steps (d) and (e) for the other side of the circle.

(g) Use the gold marker to place a tiny dot 8 cm from the centre of the circle on one of the spiral arms.

▶ **What I've learned**

Draw a diagram of your model. Include labels to explain what each part of the model represents; the black circle represents the disc-shaped Milky Way etc.

Prim-Ed Publishing ~ www.prim-ed.com   Hands-on science

Lunar craters

Earth and beyond • Earth and beyond • Earth and beyond • Earth and beyond

Objective: Determines how the height from which an object is dropped can affect the appearance of impact craters on a 'lunar' surface.

Materials
- large pan (similar to cat litter pan) or cut the bottom from a large box (include sides of at least 8 cm—box/pan can be lined with garbage bag); 'lunar surface' material such as flour; dry tempera paint; sifter or sieve; newspaper; ruler; metre ruler; 'meteorites' such as marbles, golf balls, ball bearings, pebbles etc.; plastic/material drop cloth; safety glasses/sunglasses for pupils.

▼ Motivate
- Show pupils photographs of the craters in the moon and ask pupils for their ideas of how the craters are formed. Ask if anyone has seen these craters through a telescope.
- Ask: Could there be craters on Earth? (Yes) Where are they? Are they as visible as those on the moon? (No, due to erosion and weathering or they are under the ground due to sediment covering them over time.)

▼ Experience
- Organise pupils into small groups.
- Dropping balls from different heights will allow pupils to study the relationship between the speed of the 'meteorite' with the size of the crater formed.

Safety note: Be aware that paint and flour will disperse in the air when the impact occurs. Pupils should wear safety glasses (or any type of glasses) to protect their eyes from these particles.

- Discuss the criteria for a fair test with the class. To make this activity a fair test, pupils need to decide how the 'lunar surface' will be smoothed between each drop. (The flour should not be packed tightly.)

Note: Allow some practice time for dropping 'meteorites' and resurfacing the materials in the pan before the data is recorded.

- Pupils follow the steps on the worksheet to complete the experiment and record their results in the tables.

▼ Explain
- Before pupils complete the 'What I've learned' section on the worksheet, discuss once again with the class how the height of the ball being dropped affects its speed when impacting with the lunar surface. Discuss the results with the class: How has the diameter and depth of the crater changed from the ball with the lowest speed (drop height of 30 cm) to the ball with the highest speed (drop height of 90 cm)?

Impact craters are formed when objects such as comets, asteroids and meteorites smash into the surface of the moon, the Earth, comets and other planets. The materials which were in the crater are hurled out, forming a pile of rocks around the circular hole. Scientists study craters on the moon to determine the age of different surfaces and to create a geological history of the moon. Generally, the more craters, the older that part of the moon's surface is. By recording the number, size, and amount of erosion of craters, scientists can determine some of the moon's history.

The size of the 'meteorite' and the speed it was travelling at impact are the two main factors affecting the crater's appearance. The geology of the surface material is also important. Impact craters are also found on the Earth, but due to weathering and erosion, they are more difficult to see.

Famous craters include Barringer Meteor Crater in Arizona, USA; Manicouagan and Sudbury craters in Canada; and Chicxulub in Mexico (this crater is a kilometre underground).

Parts of a crater include the bowl, the ejecta (material thrown when the impact occurred), rim, walls and rays.

▼ Apply
- Pupils construct a graph displaying the data they have recorded.
- Pupils consider other factors that could affect the appearance of a crater; for example: the mass of the 'meteorite' or the angle with which it hits the surface. Pupils design and carry out an investigation to make conclusions about the relationships between one of these factors and the appearance of impact craters. Use the science investigation framework on page xiv.

▼ Review and reflect
- Take digital photographs of the pupils completing the experiment. Pupils use the photos, plus images of impact craters from books and the Internet, to create an information poster summarising their experiment.

Lunar craters

Task: To create 'lunar craters' and demonstrate how the height that an object is dropped from affects the appearance of the crater.

▶ **What to do:**

1. Place the pan on newspaper and then add flour so it is about 3 centimetres deep. Smooth the surface, then gently tap the pan to make the flour settle evenly.
2. Sift a layer of dry paint evenly over the 'lunar' surface.
3. Choose your 'meteorite' and measure the first drop height.
4. After each drop, use your ruler to carefully measure the diameter and depth of the crater formed. Between each drop, smooth the materials and add a fresh layer of paint.
5. Record your results in the table.

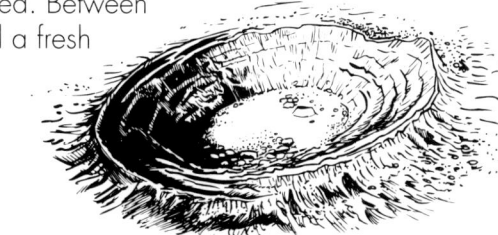

You will need

large pan or box

'lunar' surface material such as flour

dry tempera paint

sifter or sieve

newspaper

ruler

metre ruler

'meteorites' (such as marbles, pebbles, ball bearings, golf balls etc.)

safety glasses/ sunglasses

▶ **What happened?**

Drop height – 30 cm	Trial 1	Trial 2	Trial 3	Average
Crater diameter				
Crater depth				

Drop height – 60 cm	Trial 1	Trial 2	Trial 3	Average
Crater diameter				
Crater depth				

Drop height – 90 cm	Trial 1	Trial 2	Trial 3	Average
Crater diameter				
Crater depth				

▶ **What I've learned**

1. Write a sentence that summarises your findings about how the height an object is dropped from, affects the appearance of the crater it forms.

2. Draw a labelled diagram of an impact crater.

An energetic toy

Energy and change • Energy and change • Energy and change • Energy and change

Objective: Creates a toy that demonstrates how stored (potential) energy can change to action (kinetic) energy.

Materials
- For each pair of pupils: elastic band, an old-style cotton reel, sticky tape, matchstick, washer, pencil

▼ Motivate
- Show the pupils photographs or pictures of people and things displaying stored and action energy (e.g. muscle energy) and discuss how things move and where the energy for the movement comes from.

▼ Experience
- Organise the pupils into pairs and distribute the materials. The pupils can use the instructions and the diagram to help them build their 'energy changers'.
- The pupils describe what happened when they put the energy changer down on the floor—as the pencil is turned, the elastic band is twisted and tightened; when the band unwinds, the energy changer spins or 'crawls' across the floor.

▼ Explain
- Explain to the pupils that energy is something we need to make things happen. Energy can exist in many different forms. Whatever form it takes, energy can be used to make things move.
- As a class, discuss the pupils' observations and ask some volunteers why they think the energy changer moved. Pupils can then complete 'What I've learned'. The answers should indicate that the tightened elastic band is an example of stored energy. When the band unwinds, the stored energy is released as action energy, causing the energy changer to move.

▼ Apply
- Pupils draw and label a diagram (see page xvi) showing how the energy changer works.

▼ Review and reflect
- Complete a science recount on page xiii.

There are two main forms of energy:
- **action energy**—anything that is moving. Action energy is also known as kinetic energy.
- **stored energy**—anything that is stretched, compressed (squashed) or held above the ground. Stored energy is also known as potential energy.

Energy can change from one form to another but is never lost or destroyed.

An energetic toy

Task: Create a toy that demonstrates how energy can change from one form to another.

There are two main forms of energy:
- stored energy – anything that is stretched, compressed (squashed) or held above the ground.
- action energy – anything that moves.

Energy can change from one form to another but is never lost or destroyed.

You will need
elastic band
an old-style cotton reel
sticky tape
matchstick
small washer
pencil

▶ What to do:

Make an energy changer to see how one form of energy can be changed into another.

(a) Thread the elastic band through the cotton reel.

(b) Carefully snap the tip off the end of the matchstick, making sure it stays in one piece. Place it through the loop of the elastic band at one end of the cotton reel.

(c) Tape the matchstick to the cotton reel so it is unable to move.

(d) Thread the other loop of the elastic band through the washer.

(e) Place the pencil through the loop on the outside of the washer.

(f) Use the pencil to wind up the elastic band tightly.

(g) Place the energy changer on the floor.

▶ What happened?

Describe what you saw.

_____
_____
_____

▶ What I've learned

1. Explain why you think this happened. Use the phrases 'stored energy' and 'action energy' in your explanation.

   _____
   _____
   _____

2. If you could make another energy changer, what would you change to make it work more effectively?

Shoebox guitar

Energy and change • Energy and change • Energy and change • Energy and change

Objective: Makes and experiments with a shoebox guitar to explore pitch.

Materials
- For each pair of pupils: shoebox; 5 toothpicks; 4 elastic bands (of different thicknesses); pencil; scissors; glue
- If desired, a model shoebox guitar could be made prior to the lesson to show the pupils (see 'Experience').
- A real guitar

▼ Motivate
- Demonstrate to pupils the plucking of the strings of a guitar. Make sure the pupils understand that the thicker strings produce lower pitches and adjusting the tuning pegs or pressing on a fret (thereby shortening the strings) also affects the pitch.

▼ Experience
- Organise the pupils into pairs and distribute the materials. If a model guitar has already been made, it could be shown to the pupils as an example.
- Have the pupils discuss and make predictions with their partners, noting them on their worksheets. Encourage the pupils to think carefully about the demonstration of the real guitar when making their predictions.
- The pairs of pupils then make their guitars by following the given procedure and then test out their predictions. They should realise that pressing on the frets, in effect, makes the elastic band shorter and therefore produces higher sounds and that plucking a thicker elastic band will produce lower sounds than a thinner band. Some pupils may also try to vary the pitch by tightening their elastic bands around the box.
- Pupils then complete the 'What happened?' section of their worksheet.

▼ Explain
- Explain to the pupils that plucking the strings/elastic bands on their guitars causes them to vibrate, creating sound. The pitch of the sound depends on the thickness and length of the rubber band and the speed at which it vibrates. Rapid vibrations make high-pitched sounds and slower vibrations, low-pitched sounds.
- Pupils complete the 'What I've learned' section. Answers can be discussed as a class.

> The difference between high and low sounds is the rate (frequency) of the vibration. High-pitched sounds are made by rapid vibrations and low-pitched sounds are made by slower vibrations. In the case of the elastic bands, the thicker ones have more mass than the thin ones and therefore vibrate more slowly, producing a lower pitch. Also, when the elastic band is shortened (by pressing frets), it vibrates more quickly, producing a higher note.
>
> The pitch a string or elastic band vibrates at when plucked is known as its 'resonant frequency'. This is determined by the length, tension and mass per metre. Halving the length of a string or elastic band will double its resonant frequency and produce a sound eight notes higher.

▼ Apply
- Make different length 'straw oboes' from drinking straws or fill glass bottles with different amounts of liquid and blow across the top to consolidate the idea with pupils that shorter or smaller things vibrate more quickly; thereby producing higher pitches.

▼ Review and reflect
- Have the pupils make a variety of simple musical instruments, each of which can play a range of notes. The pupils can then play them for younger pupils and explain how they made the different notes.

Shoebox guitar

Task: Make a guitar from a shoebox and elastic bands to experiment with pitch.

You will need
shoebox

5 toothpicks

4 elastic bands (all of different thicknesses)

pencil

scissors

glue

▶ What will happen?

Make your pitch predictions by circling the correct word in each sentence below.

☐ Plucking a longer elastic band will produce a **higher/lower** pitch than a shorter elastic band.

☐ Plucking a thicker elastic band will produce a **higher/lower** pitch than a thinner elastic band.

▶ What to do:

Make your shoebox guitar by following the instructions below.

(a) Take the lid off the shoebox.

(b) Trace a circle of about 8 centimetres in diameter approximately 2 centimetres from the end of the lid and in the middle. Cut it out.

(c) Arrange the toothpicks in a line evenly spaced between the hole and the opposite edge of the lid to make frets. Glue them in place and replace the lid on the shoebox.

(d) When the glue is dry, slide the elastic bands lengthways over the shoebox in order from thickest to thinnest.

(e) Slide the pencil under the elastic bands and place it in the 2-centimetre gap between the hole and the end of the box to make a bridge.

▶ What happened?

1. Try out your guitar to test your predictions. Describe what you did and what happened.

   _____
   _____
   _____
   _____

2. Tick the predictions you made correctly.

▶ What I've learned

Based on your tests and results, explain the factors that determine an instrument's pitch.

_____
_____
_____
_____

Insulated flask challenge

Energy and change • Energy and change • Energy and change • Energy and change

Objective: Makes an insulated flask to demonstrate how heat energy is retained.

Materials

- For each group of pupils: hot water; 3 small identical jars with lids; 2 large jars with lids; sticky tape; newspaper; aluminium foil; bubble wrap; polystyrene foam; thermometer
- a few vacuum flasks

▼ Motivate

- Show the pupils the vacuum flasks. Tell pupils that you will shortly be challenging them to make their own flasks, so ask them to observe the containers closely. Discuss their observations. The pupils should be made aware that the inside of a vacuum flask is made of shiny material and that the inner container has a layer of insulation around it—these are the features they can try to imitate when making their own flasks.

▼ Experience

- Organise the pupils into groups and distribute the materials. They should then set one jar aside as a control and then use various materials to test out two different types of flasks. For example, they could stick aluminium foil around the jar, use shredded newspaper for insulation, place the small jar inside a large jar etc. Lids provide additional insulation. Each group could even divide in half and work on a flask each, holding a contest to see whose flask kept the water hottest after an hour. Set a time limit for the pupils to make their flasks—10–15 minutes is suggested.
- Once the flasks have been completed, an adult should pour the hot water into each of the small jars for the pupils. Screw on the lids then place them in an undisturbed area of classroom and after an hour, measure the temperatures with thermometers. They complete the 'What I've learned' section of the worksheet and discuss their ideas with the class.

Safety note: Keep pupils well away from the hot water as it is being poured into the jars and remind them to be cautious as they experiment with the jars. The water should not be so hot that it can burn.

▼ Explain

- Explain to the pupils that heat is a form of energy. Hot objects have higher internal levels of energy than cold objects. Heat can have several effects on materials—such as increasing their temperature or pressure or causing expansion, melting or vaporisation. Heat can pass from place to place or through objects by conduction, convection or radiation. Conduction transfers heat through an object or substance (e.g. a heated frying pan causing food to cook); convection transfers heat by circulating a heated liquid or gas (e.g. a heater warming the air in a room); and radiation transfers heat through space (e.g. the sun warming the earth).
- In this activity, the pupils will be trying to make a container that prevents loss of heat through conduction and radiation.

> A commercial vacuum flask can prevent the loss or entry of heat through conduction and convection, as well as radiation. The inner container is usually made from glass or stainless steel. Most of the air between the inner and outer containers is removed to create a partial vacuum (blocking convection because there are hardly any air molecules to carry heat); and the surface of the inner container is reflective (reflecting heat and preventing radiation). In addition, vacuum flasks have lids made from materials which do not conduct heat well. Because heat cannot pass into the inner container, a vacuum flask can also keep cold liquids cool.

▼ Apply

- Pupils experiment with their vacuum flasks with ice cubes, tracking how long the ice takes to melt.

▼ Review and reflect

- Complete a labelled scientific diagram of a commercial vacuum flask on page xvi.

Insulated flask challenge

Task: To make an insulated flask to demonstrate retention of heat energy.

An insulated flask like a thermos can help hot liquids retain heat. Try making your own insulated flask.

You will need

3 small identical jars with lids

2 large jars with lids

sticky tape

newspaper

aluminium foil

bubble wrap

polystyrene foam

thermometer

▶ What to do:

1. Look at an insulated flask, like a thermos. List some of its features in the space below.

2. Put one of the three small jars aside. This will be your 'control'. Use it to show how quickly water cools down in your classroom environment with no insulation. The other two jars are yours to experiment with. Use some or all of the materials to make the most effective flasks you can. Try out a different idea on each jar.

3. Draw labelled diagrams of your two insulated flasks.

   FLASK 1

   FLASK 2

4. Tick the one you think will keep the water hot for the greatest amount of time.

5. Ask an adult to pour the same amount of hot water into each of the three small jars. **Safety note:** Be careful when working with hot water.

6. After an hour, measure the temperature of the water in all three jars. Write your results below.

   Control: _____   Flask 1: _____   Flask 2: _____

▶ What happened?

Which flask kept the water hottest? Why do you think this was?

▶ What I've learned

Suggest some ways in which you think you could have improved your best insulated flask.

Skill tester game

Energy and change • Energy and change • Energy and change • Energy and change

Objective: Makes a simple game that uses a closed electrical circuit.

Materials

- For each group of pupils: shoebox; electrical tape; 15-cm length of flexible bare wire; 80-cm length of flexible bare wire; 3-cm length of insulated wire with both ends bared; 50-cm length of insulated wire with both ends bared; wire clippers; pen; 1.5 volt light bulb in a socket; scissors; AA battery
- A small buzzer could be substituted for the light bulb in this activity; however, bulbs are more readily available (Note: When making a circuit, the device and battery should be matched; e.g. if the bulb is 1.5 volt, it needs a 1.5 volt battery)

▼ Motivate

- Demonstrate a closed circuit to the class using a battery, wires and a light bulb. Discuss how electricity travels from a power source, around a circuit and back to the power source. The following diagram could be drawn on the board.

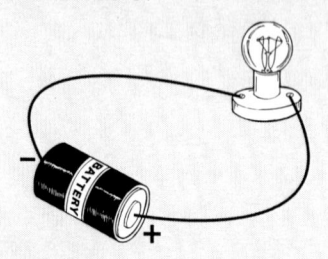

▼ Experience

- Organise the pupils into groups and distribute the materials. It is recommended that a sample skill tester game is made prior to the lesson to show pupils how to construct the game.

Safety note: Take time to discuss the dangers of mains electricity. All work and equipment for this activity should be supervised as even low-voltage batteries have the capacity to burn or start a fire. Adult help should also be provided with the wire clippers.

- When pupils are making their games, all the connections to the battery and the wires need to be very firmly taped. If the game does not work, pupils should check all connections.
- Some pupils may be confused as to why the game demonstrates a closed circuit when the loop is not attached to the bulb socket. Encourage the pupils to think carefully as to why the game might work—i.e. the loop touching the bent wire completes the circuit.

Note: Holes A and C may need to be made bigger by using a felt pen or by wiggling a pen around.

- Once the groups have tested their games, the questions can be answered as a group and then discussed with the class.

▼ Explain

- Explain to the pupils that electricity travels from a power source, such as a battery, around a circuit (a series of conductors) and back to the power source. No electricity will flow if there is a gap in the circuit—this is the reason why the pupils had to check all the connections carefully for their game to work.

> An electric current is a flow of microscopic particles called electrons through wires and electric components. As water is pushed through pipes by a pump, an electric current is pushed through wires by a battery. An electron has a negative charge. The battery has a negative terminal and a positive terminal. The negative terminal of a battery will push negative electrons along a wire and the positive terminal of a battery will attract negative electrons along a wire. In this activity, an electric current flows from the negative terminal of a battery, through the light bulb to the positive terminal.

▼ Apply

- Have the pupils use the concept of a closed electrical circuit and the game they made to help them design a working burglar alarm for their bedroom door.

▼ Review and reflect

- Pupils draw a labelled scientific diagram (page xvi) of the skill tester game they made.

Skill tester game

Task: To make a skill testing game using an electrical circuit.

What to do:

1. Push the pen into the shoebox to make four holes—A, B, C and D—as shown in Figure 1.

2. Tape one end of the 3-cm length of insulated wire to one end of the battery. Tape the other end of the wire to one side of the socket.

3. Tape one end of the 50-cm length of insulated wire to the free end of the battery.

4. Tape one end of the 80-cm length of bare wire to the free side of the socket.

5. Make a small loop from the 15-cm length of bare wire. You may need to use the wire clippers to help you. Join the loop to the free end of the 50-cm length of insulated wire and seal it with electrical tape.

6. Check Figure 2 to ensure that all your connections are correct. Add extra tape to make sure everything is secure.

7. Now you are ready to put your game together! Take the lid off the box. Put the battery, the socket and everything you have connected to them inside the box.

8. Poke the light bulb through hole A and into the socket. Poke the 80-cm length of wire up through hole B and shape it into a series of bends. Poke the loop through hole C and thread it onto the 80-cm length of wire. Push the end of the 80-cm length of wire into hole D and bend it securely inside of the lid so it doesn't slip.

9. Carefully put the lid of the shoebox back on. It should look like Figure 3. Play your skill tester game by trying to move the loop over the bent wire without touching it. How skilful are you?

You will need

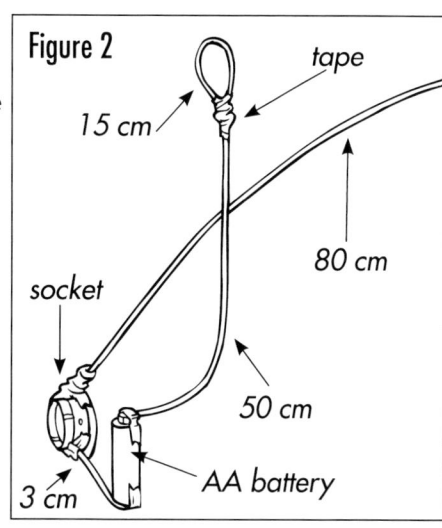

- shoebox
- electrical tape
- 15-cm length of flexible bare wire
- 80-cm length of flexible bare wire
- 3-cm length of insulated wire with both ends bared
- 50-cm length of insulated wire with both ends bared
- wire clippers
- pen
- 1.5 volt light bulb in a socket
- scissors
- AA battery

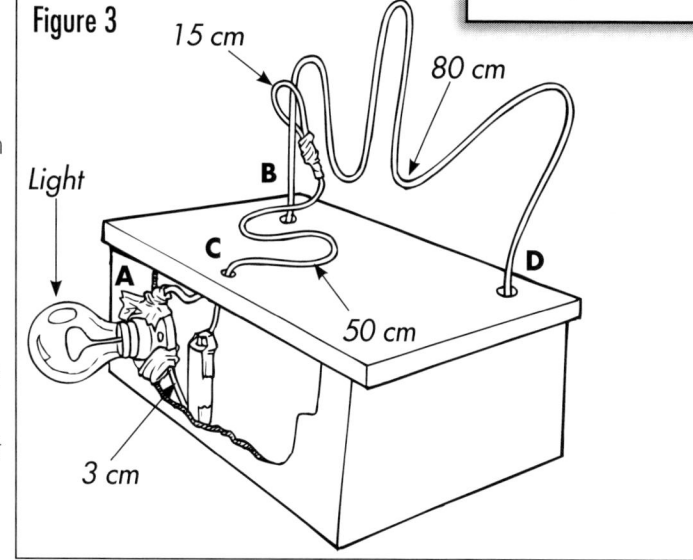

What happened?

1. What did or should have happened when the loop touched the wire?

_____

2. Did your game work? Explain why or why not.

_____

What I've learned

This game is an example of a _____ electrical circuit.

Prim-Ed Publishing ~ www.prim-ed.com   Hands-on science

An electric magnet

Energy and change • Energy and change • Energy and change • Energy and change

Objective: Makes and tests an electromagnet.

Materials

- For each group of pupils: 50-cm length of insulated wire with both ends bared; assorted nails; AA battery; battery holder (or electrical tape); paperclips; extra wire (wires with alligator clips attached can also be used).

▼ Motivate

- Remind pupils how an electrical circuit is created (see page 26).
- Discuss with pupils magnetism and what types of materials are magnetic. (some metals, including iron) Demonstrate different types of magnets, then ask pupils if they think it is possible for a magnet to be turned on and off.

▼ Experience

- Organise the pupils into groups and distribute the materials. Pupils follow worksheet to make their first electromagnet. When they understand how it works, pupils can experiment with different materials. They can try the three suggestions on the page plus one variable of their own choosing; e.g. a tighter coil, a looser coil, shorter wire, a thinner nail. The paperclips should be linked together by magnetism to test the strength of their electromagnets.

Safety notes:

- Take time to discuss the dangers of mains electricity. All work and equipment for this activity should be supervised as even low-voltage batteries have the capacity to burn or start a fire.
- Warn pupils that if the battery becomes hot while conducting this experiment, they should disconnect the circuit (particularly if electrical tape is being used in place of a battery holder).

- When the pupils are happy with their final electromagnet design, a class challenge can be held. The features of the winning design(s) should be discussed before the pupils answer the final question on the page.

▼ Explain

- Explain to pupils that when an electric current flows through a wire, it creates a magnetic field. This field can magnetise any metals which contain iron (such as a nail). The magnetic field in this experiment is concentrated because the wire is coiled around the nail.

> In 1820, Danish physicist Hans Ørsted was the first person to note that an electric current creates a magnetic field.
>
> Most electromagnets are made by coiling wire around an iron core. Electromagnets only remain magnetic while the electric current is flowing. Electromagnets are widely used in devices like doorbells, electric guitars, motors and generators and are used in industry for moving scrap iron and steel.

▼ Apply

- Have the pupils compare the strength of their electromagnet with that of other household magnets.

▼ Review and reflect

- Write a science procedure for making and testing an electromagnet and display it as a poster.

An electric magnet

Task: To make an electromagnet using an electric circuit.

An electromagnet is a special type of magnet. It only became a magnet through electricity. How? When an electric current flows through a wire, it creates a magnetic field. If the wire is attached to a piece of metal containing iron, the metal becomes magnetic when the electric current is flowing through it.

You will need

- 50-cm length of insulated wire with both ends bared
- assorted nails
- AA battery
- battery holder
- paperclips

▶ What to do

Try making and testing your own electromagnet.

(a) Coil the wire firmly around the nail, leaving about 10 cm free at both ends.

(b) Use the loose ends of the wire and the battery to create a closed circuit. The nail should have turned into an electromagnet!

(c) Try out your electromagnet. How many paperclips can it pick up? _____

Electromagnet challenge!

1. With your group, try making the strongest electromagnet you can by changing some of its features. Here are some suggestions you could consider:

   a longer nail   *a thicker nail*   *longer wire*

   Try four different things. Make sure you only try one change at a time! Each time, write how many paperclips your new electromagnet design picks up.

Change made to original	Number of paperclips

2. Draw and label your electromagnet that collected the most paperclips.

3. Test your best electromagnet against those of other groups. The winning electromagnet picked up ☐ paperclips. On the back of this sheet, explain why you think this design was the winner.

Moving by force

Energy and change • Energy and change • Energy and change • Energy and change

Objective: Makes a toy that demonstrates the forces of gravity and friction.

Materials

- For each pupil: drinking straw; elastic band; white, thin card; glue; scissors; sticky tape; template (see pupil page)

▼ Motivate

- Discuss the meaning of the word 'force'. Explain that, in a scientific sense, a force is something we experience all the time but cannot see. Forces act on objects to make them move, change shape or direction, speed up or slow down. Two types of force are gravity and friction.
- Ask pupils what they understand by the word 'gravity'. Explain that gravity is a force that draws all things on the Earth down towards its centre. Gravity is the reason why we don't fall off the Earth and why, if you drop an object, it falls to the ground.
- Ask pupils what they understand by the word 'friction'. Explain that friction is a force that resists motion. It occurs because of bumps on the surfaces of two objects. The pupils can experience friction by trying to slide an eraser across a desk.

▼ Experience

- Distribute the materials to the pupils and have them follow the instructions to make the toy. If the template is cut out correctly from the card, the template should look like this:

Pupils can decorate the bird template if they wish. The piece of straw does not need to be placed precisely in the centre of the crease; however, the two bird shapes do need to be stuck together for the toy to work.

Safety note: Explain to pupils that they need to keep the elastic band well away from their faces as they use their toy. Safety goggles could be used to protect their eyes.

- If the toy has been constructed correctly, the bird should 'peck' continuously as it moves down the elastic band. If this does not occur, the pupils can try making the straw shorter or cutting it to make it thinner.

▼ Explain

- Explain that the toy bird pecks as it moves down the elastic band because the bird shape causes the straw to twist, making it dig into the rubber band. The friction this creates causes the bird to stop. However, because gravity is acting on the bird, pushing it down, it drops slightly before the straw digs into the rubber band again, causing the pecking motion. This toy could also be used to help explain the conversion of potential energy to kinetic energy.
- Pupils complete the 'What I've learned' section of the worksheet.

> Gravity is a force of attraction that acts between two objects. The greater the mass of an object, the more gravitational pull it exerts; therefore, the object with the greater mass will pull the smaller object towards it. Gravity is what causes everything on or near the Earth to be pulled downward. In the late 1600s, Sir Isaac Newton was the first to define the law of gravity, after he famously saw an apple fall from a tree. In 1915, Albert Einstein expanded upon Newton's theory when he published his theory of relativity.
>
> The force that occurs when two objects or surfaces are moving against each other is called friction. Friction is a force that resists motion. It always acts in the opposite direction to the motion and occurs because of the bumps on the surfaces of two objects. Even the smoothest of surfaces have these little bumps! When two objects slide together and there is friction, they lose energy and slow down or stop moving. The energy changes from kinetic energy (moving energy) to heat energy. This is why when we rub our hands together, we feel heat.

▼ Apply

- Have the pupils experiment with different weights and elastic bands to make the most effective pecking toy they can.

▼ Review and reflect

- Pupils could demonstrate the toy to younger pupils and give a simple explanation of how it works.

Moving by force

Task: To make a moving toy that demonstrates gravity and friction.

▶ What to do:

1. Fold the piece of card in half. Cut out the template at the bottom of the page and stick it to the card so the solid line on the shape is lined up against the crease in the card.

2. Cut out the shape, cutting through both layers of card. Unfold the card shape you made. You should now have two bird shapes.

3. Cut a 3-cm piece of straw and tape it to the crease between the two bird shapes. Fold the card back, so the straw is on the outside, and then glue the inside of the two bird shapes together.

4. Cut the elastic band so it is now a ribbon of band and thread it through the straw. Slide the bird shape to the top of the length of elastic band.

You will need

- drinking straw
- elastic band
- thin, white card
- glue
- scissors
- sticky tape
- template (see below)

▶ What will happen?

If you were to stretch the elastic band vertically, what do you think would happen to the bird?

_____

_____

▶ What happened?

Test your toy by holding each end of the length of elastic band. Stretch the elastic band vertically, shake your hands a little, then hold them still, keeping the elastic band stretched tightly. Write what happens.

Safety note: Keep the elastic band away from your face in case it snaps.

▶ What I've learned

Explain how you think each of these forces affected the movement of your toy bird.

- Gravity: _____

_____

_____

- Friction: _____

_____

_____

Prim-Ed Publishing ~ www.prim-ed.com   Hands-on science   **31**

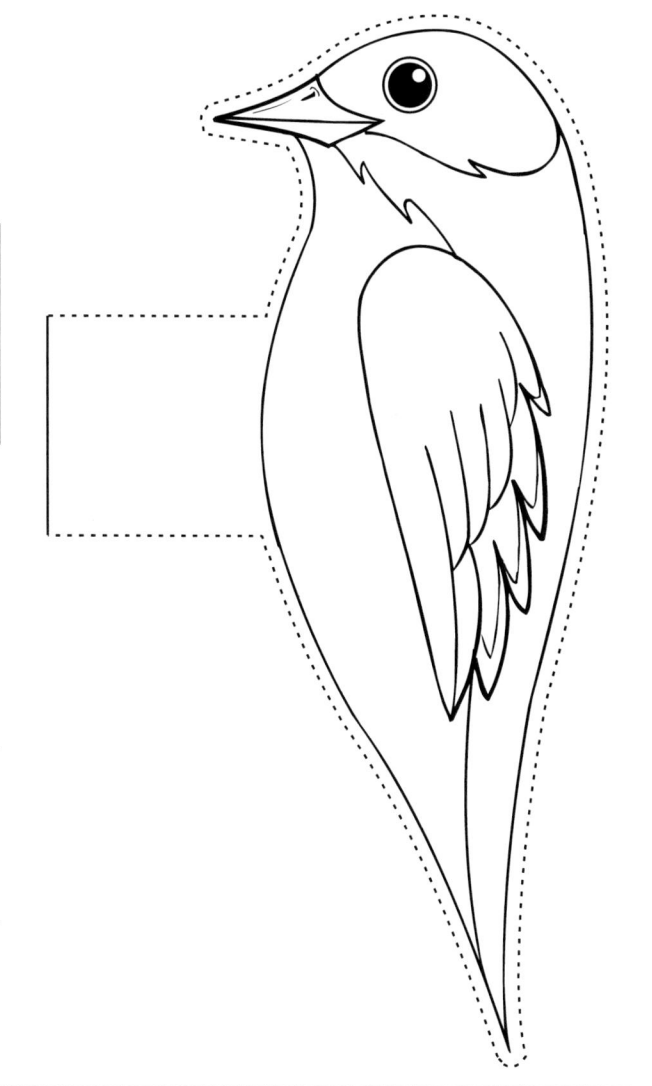

Balloon rocket

Energy and change • Energy and change • Energy and change • Energy and change

Objective: Makes a balloon rocket to demonstrate thrust.

Materials

- For each group of pupils: 5 balloons (of various shapes, including a long, thin balloon); clothes pegs; thick felt pen; 5-metre length of string; drinking straws; sticky tape; scissors; thin card; paperclips; tape measure

▼ Motivate

- Blow up a balloon and let go of it in the classroom. Ask the pupils why they think it moved in the way it did. Discuss how rockets move in the same way due to a pushing force called 'thrust'. Read with pupils the information at the top of the worksheet to help make the concept clearer. Display some pictures of rockets and discuss their physical features—their long, thin shape; nose cone and fins. This information will provide the pupils with ideas when they experiment with their balloon rockets.

▼ Experience

- Organise the pupils into groups and distribute the materials. Each group of pupils will need to work in a separate area of the classroom, with enough room for them to stretch out their string. The pupils can then follow the instructions to test their five balloons and answer the questions when they have finished. They should discover that the long, thin balloon made the best rocket. (Note: a smaller long, thin balloon will travel further than a large long, thin balloon because it has more air pressure and less mass).

- The pupils can then try modifying their best balloon rocket design to travel even further. They can use the card and also modify other variables (e.g. adding more air to the balloon, making the straw longer). The pupils should discover that the most effective ways to make the rocket travel further are by adding a nose cone and fins and beginning with more air in the balloon. A longer string could be rigged up across the whole classroom for the final test.

▼ Explain

- Explain to the pupils that a balloon rocket moves forward as air is forced out of the balloon's opening, creating thrust. Rockets work on the same principle—only their thrust is created by gases forced out of the rocket and they are sent into the sky, rather than along a string.

- Pupils complete the 'What I've learned' section of the worksheet.

> Sir Isaac Newton's third law of motion states that 'every action has an equal and opposite reaction'. In the case of the balloon rocket, the action is the compressed air escaping out of the opening of the balloon and the opposite reaction is the forward movement of the balloon.

▼ Apply

- Have the pupils design and make a balloon car, using the same principles as for the balloon rocket.

▼ Review and reflect

- Pupils write a science recount on page xiii.

Balloon rocket

Task: To make a balloon rocket.

A rocket can take off into the sky when burning fuel is blasted out of the back of the rocket. This forces the rocket to move upward. The force that drives the rocket upwards is called 'thrust'.

▶ What to do:

Make your own rocket using a balloon.

(a) Label each of your balloons with a number from 1 to 5.

(b) Thread the string through a drinking straw. You may need to cut the drinking straw if it is too long for the balloon.

(c) Blow up Balloon 1 and twist its opening. Seal it with a clothes peg. Use sticky tape to attach the balloon to the straw.

(d) Have two members of your group hold either end of the string and pull it tight. Slide the balloon to the end of the string.

(e) When you are ready, release the clothes peg. Measure the distance the balloon rocket travels and write it on the balloon.

(f) Test the four other balloons using the same procedure.

You will need

5 balloons
clothes pegs
thick marker pen
5-metre length of string
drinking straws
sticky tape
scissors
thin card
paperclips
tape measure

▶ What happened?

1. Describe the size and shape of the balloon which travelled the least distance.

   _____

2. Describe the size and shape of the balloon which travelled the greatest distance.

   _____

3. Try to make your most successful balloon rocket travel even further. You may use the card and/or the paperclips. Note what you tried and whether it had any effect.

4. Test your balloon rocket against those of other groups.

▶ What I've learned

Describe how you think a balloon rocket works and what makes a balloon rocket work better.

_____

_____

_____

_____

Up periscope!

Energy and change • Energy and change • Energy and change • Energy and change

Objective: Makes a periscope to explore reflection.

Materials

- For each pair of pupils: 2 one-litre milk cartons; 2 small, square-shaped mirrors (these mirrors do not need to be identical but do need to fit inside a carton—old face powder compact mirrors would be suitable); scissors; ruler; masking tape; pencil; extra card; protractor
- Before the lesson, make a periscope using only one milk carton for pupils to compare to the version they will make during the activity.

▼ Motivate

- Discuss what the pupils understand by the word 'reflection'. What sorts of materials can we see our reflection in? Discuss how reflections occur—light is reflected off a surface, such as a mirror, at exactly the same angle as it hits the surface.
- Have the pupils view their own reflections in a mirror. They should move the mirror closer and then further away from their face. What happens to the image? (The further away the mirror, the smaller the reflection).
- Using two small mirrors, investigate how mirrors held at different angles produce different images (i.e. mirrors held at 90° give two images; mirrors held at 60° give three images; parallel mirrors give infinite images; and mirrors held apart, but angled to face each other in opposing corners, act like a periscope). Ask if pupils know what a periscope is and where they are used.

▼ Experience

- Organise the pupils into pairs and distribute the materials. Explain to the pupils that they will need to leave a couple of centimetres from the corner of the milk carton when cutting the windows and the slots. Some pupils may also require help measuring the 45° angles. Once they have constructed their periscopes, have the pupils test that they work. If images cannot be seen or are crooked, they will need to adjust one of the mirrors. The pupils can then answer Question 1. The shorter model periscope (see 'Materials') can then be viewed by the pairs. The pupils should notice that images appear to be bigger than in their longer periscope.

Safety note: Mirrors should be handled with care as their edges can be sharp and they may break if dropped.

▼ Explain

- Explain to the pupils that periscopes are used to see objects from a distance or when normal vision is obstructed. Periscopes work because the mirrors are angled at 45°. Light hits the top mirror at 45°, and is reflected to the bottom mirror at 45°, then reflects the original image at 45° straight into the viewer's eye.

Periscopes are used not only in submarines and tanks but also in scientific laboratories and even nuclear reactors. The longer the periscope, the smaller the image that will appear. For this reason, periscopes in submarines and tanks have magnifying lenses placed between their mirrors.

▼ Apply

- Ask pupils to try designing and making a periscope that uses more than two mirrors. What purpose could it serve?

▼ Review and reflect

- Build a cardboard submarine that has a periscope for a younger class at the school. The pupils can explain to the younger pupils how the periscope works.

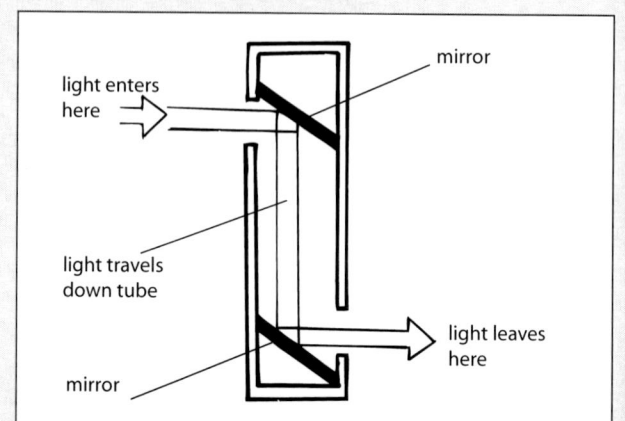

The mirrors must be at 45° angle and line up, otherwise the image will bounce off the first mirror into the side of the periscope and will not reach the second mirror (or the viewer's eye).

Up periscope!

Task: To make a periscope.

▶ What to do:

1. Cut the peaked top off both milk cartons. Tape the cartons together end to end.
2. Cut two square windows out on opposite sides of the milk cartons—one at the top and one at the bottom—as shown in the picture.
3. Use a protractor to help you measure and cut two diagonal slots at a 45° angle in one side of your box, as shown in the picture. Do not cut right to the edge of the carton.
4. Trace the square base of the milk carton onto a piece of card. Add a 2-centimetre tab to one side. Repeat so you have two identical pieces of card.
5. Tape each of your mirrors onto a piece of card. Slide a piece of card into each of the slots you made, (reflective sides facing each other) leaving the 2-centimetre flap sticking out of the side of the carton. The mirrors should fit snugly inside the cartons.

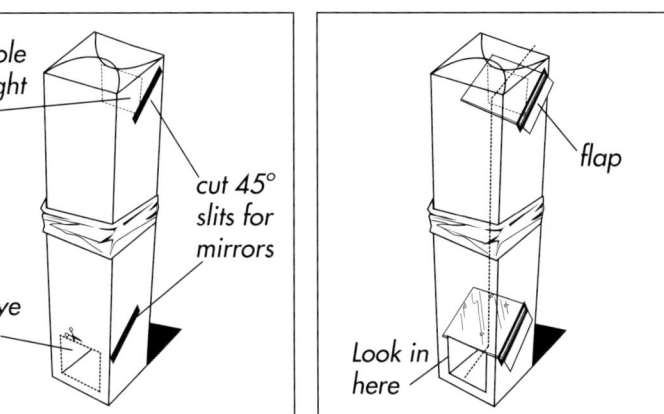

You will need

2 one-litre milk cartons
2 small mirrors
scissors
ruler
masking tape
pencil
extra card
protractor

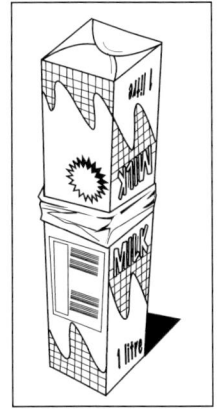

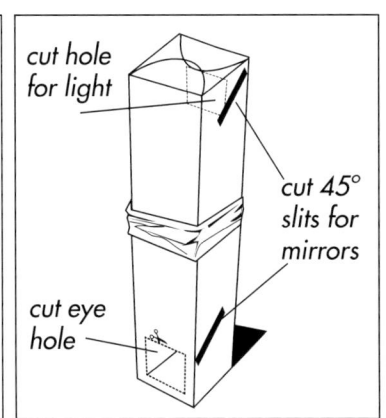

▶ What happened?

1. Use your periscope to see around a corner, over a desk or up onto a shelf by looking through one of the windows. Describe the image you see in the periscope.

_____
_____
_____

2. Look through a shorter periscope. What is the difference?

_____
_____

▶ What I've learned

Describe how you think a periscope works.

_____
_____
_____
_____

A shadowy theatre

Energy and change • Energy and change • Energy and change • Energy and change

Objective: Understands how the distance and angle of a light source affects an object's shadow.

Materials

- For each group of pupils: shoebox; sheet of white A4 paper; sticky tape; craftsticks; torch; scissors; various transparent, translucent and opaque materials – card, coloured cellophane™, fabrics, drinking straws etc.

▼ Motivate

- Ask pupils what they think is required to cast the shadow of an object. (a light source) Using a torch, projector or other light source, create hand shadows on a wall or other light-coloured surface. Investigate the different shapes and images that can be made.

▼ Experience

- Organise the pupils into groups and distribute the materials. The pupils can then make their puppet theatres and think up a simple story-line or use a story they are familiar with (e.g. a fairytale, a story covered in class). The pupils can use the torch to cast shadows of different materials onto the screen of their puppet theatre. They should discover that figures closer to the screen (further away from the torch) will create bigger, sharper shadows than those further away (closer to the torch). Coloured transparent materials, such as cellophane™, will produce coloured shadows. The pupils may also move the torch to create different shadows; e.g: when the torch beam is at a very high or very low angle in relation to the object, it will produce longer shadows.

- When the pupils have explored the shadows they can make, they can construct their characters. Some pupils may also like to decorate their 'stage sets' with scenery or props. When the pupils rehearse their plays, they may or may not choose to use voices. Groups will need to have someone watch their rehearsals to make sure their play is effective. When groups perform their plays, the puppet theatres could be set on a low bookshelf or other piece of furniture for the 'puppeteers' to hide behind. Pupils may also like to write stage directions for their plays so each member of the group knows when and where to move the different characters.

- After all the plays have been performed, the pupils complete the 'What happened?' section on the worksheet. The answers can then be discussed as a class.

▼ Explain

- Explain to the pupils that light travels in straight lines. Any object which blocks the source of light produces an area of shade behind it and casts a shadow on any surface in that area. If the light source is smaller than the object, the shadow will be uniformly dark.

- Pupils complete the 'What I've learned' section of the worksheet.

> Considering the sun as a giant light source, shadows are longest early morning and late afternoon and shortest at midday. The length of a shadow changes as the sun appears at different angles in the sky. At midday, when the sun is almost directly overhead, there is very little shadow. The position of the sun changes as the Earth rotates on its axis. In bright sun, objects have dark, clearly defined shadows. On cloudy days, shadows are lighter and more blurred.

▼ Apply

- Have the pupils track their shadows outdoors over the course of a day, considering the sun as a giant light source.

▼ Review and reflect

- Pupils draw and label the different shadows they made during their plays. The labels should explain the distance and angle of the torch in relation to each puppet.

A shadowy theatre

Task: To experiment with a light source to create shadow puppets.

▶ What to do:

1. To make your puppet theatre, cut out the base of the shoebox. Cover and tape the sheet of paper over the hole in the box to make a screen.

2. With your group, think of a simple story with three or four characters you can perform as a shadow puppet play. Your play should be around five minutes in length.

3. When you have decided on your characters, switch on the torch and shine the beam onto the back of the screen to make it light up. Experiment with different materials to create your puppet characters.

4. When your group is ready, make your puppets. Stick each puppet to a craftstick to act as a handle.

5. Practise your play, moving the torch and the puppets to create different shadow effects to make your play interesting.

You will need

shoebox

sheet of white A4 paper

sticky tape

craft sticks

torch

scissors

various transparent, translucent and opaque materials—e.g. card, coloured cellophane™, fabrics, drinking straws

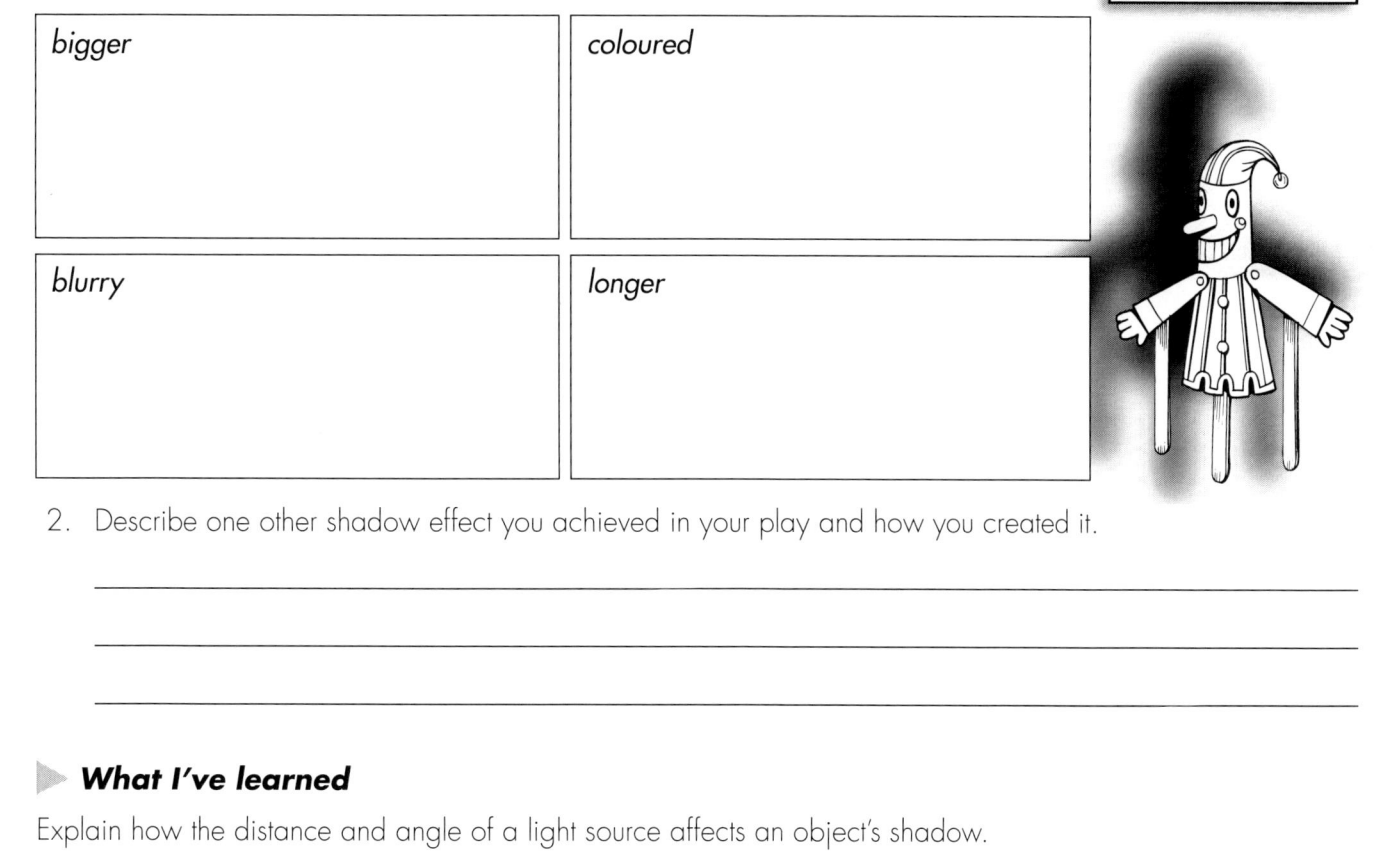

▶ What happened?

1. After you have performed your play for the class, describe what you needed to do to make shadows:

bigger	coloured
blurry	longer

2. Describe one other shadow effect you achieved in your play and how you created it.

_____

_____

_____

▶ What I've learned

Explain how the distance and angle of a light source affects an object's shadow.

_____

_____

_____

Prim-Ed Publishing ~ www.prim-ed.com   Hands-on science   **37**

Micro-organisms in food production

Life and living • Life and living • Life and living • Life and living • Life and living • Life and living

Objectives:
- Classifies foods by the type of microbe used in its production.
- Creates an informative display illustrating the types of micro-organisms used.

Materials
- A wide collection of cleaned food wrappers and containers from foods which require micro-organisms for their production (yoghurt, cheese, bread, salami, soy sauce etc.), large backing card, glue, scissors

▼ Motivate
- Ask pupils: What are micro-organisms? What do we know about them? Are they all bad? How are they used in the food industry?

▼ Experience
- Ask pupils to list all the foods they know which require micro-organisms for their production, e.g. yeast in bread, mould in some cheeses, bacteria in cheese and yoghurt. From Internet research, pupils can add more foods and record the names of the more commonly used microbes.
- The logos created by the pupils do not necessarily have to relate to what each microbe on food looks like but, maybe, how each behaves; e.g. bubbles for yeast.
- Working with the logos, slogans and food wrappers, pupils create a montage providing information about the three types of micro-organisms and the food processes that each are used in.

▼ Explain
- Pupils complete the 'What I've learned' section of the worksheet.

> - Yeast is a fungus that causes fermentation; the production of carbon dioxide and alcohol when in the presence of sugar. The carbon dioxide gas causes the dough to rise. Close examination of a slice of bread will reveal small air holes made by the carbon dioxide.
> - Bacteria are required in the production of many dairy products. In cheese-making, fresh milk has to be ripened by bacteria which convert the milk sugar (lactose) to acid and coagulate the protein (casein). This produces the curds and whey. The whey is drained and the curds salted and pressed into moulds to be matured.
> - Mould-ripened cheeses have specific moulds added to them at the dried curd stage before pressing.

> - Yoghurt production also requires the addition of bacteria which convert the milk sugar to acid and coagulate the protein.
> - During the active growth phase of the micro-organisms, temperature is maintained to promote optimum growth. When the required level of growth is reached, the food is refrigerated, preventing further fermentation.

▼ Apply
- On the Internet, pupils find a simple recipe for making yoghurt. They record the procedure and comment on the final product.

▼ Review and reflect
- Pupils complete a science reflection on page xi.
- Pupils write a report outlining what they have learned from the activity. Pupils should consider how it has, if at all, affected their opinion of micro-organisms. They suggest ideas for further research of the topic.

Micro-organisms in food production

Task: To make an informative montage using food wrappers, containers, slogans and logos.

▶ What to do:

1. For each food example, conduct research on the Internet to determine what type of micro-organism is used in the production of that food.
2. Classify each wrapper/container into groups based on the type of micro-organism used in that food's production.
3. Use the table to record the foods that use each micro-organism.

bacteria	yeast	mould

4. Choose one product from each column and research on the Internet to determine what role the micro-organism plays in the production process of that food.

bacteria	
yeast	
mould	

5. Design a logo and write slogans to use in your collage promoting each mico-organism.

bacteria	yeast	mould	bacteria
			yeast
			mould

6. Make and display your collage.

▶ What I've learned

On the back of this sheet, write an explanation describing how micro-organisms are used in the food industry.

You will need

a wide selection of cleaned food wrappers and containers

large backing card

glue

scissors

Balloon pump

Life and living • Life and living • Life and living • Life and living • Life and living • Life and living

Objectives:
- Determines the importance of sugar in the fermentation process of food.
- Determines the optimum temperature for fermentation.
- Recognises that yeast is a living organism.

Materials
- labels, black marker, packets of dried yeast, sugar, cold water, tap-water, kettle, thermometer, teaspoon measure, cup measure, 6 small plastic bottles, funnel, 6 small balloons

▼ Motivate
- Ask: Look at dried yeast. Describe it. What does it do? What is it used for?
- Yeast and sugar can be used together to create the most important reaction (fermentation) in at least two popular food products, what are they? (baked goods and alcohol)
- Ask: What conditions might affect how well the yeast and sugar react? (temperature)

▼ Experience
- In groups, pupils have to consider how a fair test may be set to determine how sugar effects fermentation and if reaction is dependent on temperature.
- Explain to pupils that, to activate the dried yeast, water is required. Cold (straight from the fridge), warm (~30 °C) and boiling water can be used to determine the effect of temperature on yeast growth.
- Outline to pupils that the following samples should be made up:

	Cold water	Warm water	Boiling water
Yeast only			
Yeast and sugar			

- The exact temperature of the water is not crucial, as long as it is the same for each sample (yeast only, yeast and sugar). The same amount of yeast and of sugar should be used for each sample.
- Suggest to the pupils that adding the sugar to the balloon, and not directly to the yeast, will ensure that any reaction, when the sugar and yeast are combined, is contained within the closed environment when the balloons are attached to the tops of the bottles.
- Pupils plan the tests, complete the table and conduct the tests as they have suggested. Pupils complete the 'What happened?' section of the worksheet.

Safety note: Ensure pupils are aware of dangers associated with hot water. If required, use an adult helper or pour the boiling water for the pupils yourself.

▼ Explain
- The carbon dioxide produced as the yeast reacts with the sugar causes the balloon to inflate. The stronger the reaction, the more the balloon inflates.

> Depending on the strain of yeast, about 30 °C is the optimum temperature for fermentation to occur and will provide the best reaction. The yeast will still react with the sugar at a cold temperature but it will be much slower. There will be no reaction with the yeast and sugar using boiling water as the yeast is destroyed at that temperature.
>
> Where yeast is used in the food industry, optimum conditions for fermentation are observed in order to provide the most effective reaction.

▼ Apply
- Pupils determine the optimum ratio of yeast to sugar for fermentation.

▼ Review and reflect
- Pupils complete a science report on page xii.

▼ Answers
What I've learned
1. Teacher check
2. Yeast is a living organism.

Balloon pump

Task: To determine the importance of sugar and temperature in the fermentation process of food.

You will need

labels
black marker
packet of dried yeast
sugar
cold water
tap-water
kettle
thermometer
teaspoon measure
cup measure
6 small plastic bottles
funnel
6 small balloons

▶ What to do:

1. In your group, plan and outline how you would conduct a fair test to show that sugar and temperature are essential for fermentation. Use the table below.

aim: (Why you are doing the test.)	
method: (How will you do the test?)	
fair test: (How will you ensure the test is fair?)	

2. Proceed with the test using your plan.

▶ What happened?

Write what was in each bottle and sketch what happened.

1.	2.	3.	4.	5.	6.

▶ What I've learned

1. What do the results tell you about the best conditions for yeast fermentation?

2. Reorganise the letters and rewrite the sentence.

   taeys si a iigvnl migsaron.

Prim-Ed Publishing ~ www.prim-ed.com   *Hands-on science*

Masters of microbes

Life and living • Life and living • Life and living • Life and living • Life and living • Life and living

Objective: Learns about some of the major breakthroughs in medical history.

Materials
- Internet access, library, encyclopaedias, medical history reference books

▼ Motivate
- Ask: What do you think is the smallest living thing?
- Ask: What personal experience do we have of micro-organisms? (infections, tooth decay, rotting food, yoghurt, veined cheeses, bread)
- Ask: How can we fight pathogenic (illness causing) micro-organisms?
- Ask: Why are we lucky living in an era where many diseases have been eradicated or, at least, controlled?

▼ Experience
- Pupils use the worksheet to read information about van Leeuwenhoek, Jenner, Pasteur and Lister. Discuss the value of each person's discovery.
- Discuss with pupils the importance of current vaccines and antibiotics.
- Pupils complete the worksheet.

▼ Explain

Microbiology is the study of micro-organisms. Micro-organisms are so small that, individually, they cannot be seen by the naked eye. Since their discovery by van Leeuwenhoek, many have been identified and the conditions for their optimum growth identified. Once the specific microbe responsible for a disease has been isolated, finding a cure for the disease becomes possible.

There are many kinds of micro-organisms; bacteria, viruses, yeasts and moulds. Although there are many pathogens and organisms that spoil food, not all microbes are bad. Those found naturally in the gut are essential for balance in the digestive tract. Antibiotics upset this balance because they also destroy all the good bacteria.

Food poisoning often occurs even when the food looks perfectly acceptable. This is because pathogens and food-spoilage organisms are not the same. Food-spoilage organisms are easily killed by adequate cooking procedures, while many pathogens are able to resist unsuitable growing conditions (cooking and freezing) by forming protective spores. When optimum growing conditions return, the pathogens return to their normal state and continue to multiply rapidly.

▼ Apply
- Pupils research to write a science report (page xii) about one of the medical discoveries mentioned.

▼ Review and reflect
- Pupils complete a 'Before and after' chart on page xvii.
- Pupils research living conditions in towns and cities during the eras shown on their time lines. How might these conditions have contributed to the short life expectancy of those times?

▼ Answers

1. Teacher check
2. Alexander Fleming – discovery of penicillin

   Louis Pasteur – discovery of rabies vaccine

   Walter Reed – transmission of yellow fever by mosquitoes

   Jonas Salk – discovery of polio vaccine

   Selman Waksman – discovery of an antibiotic effective against tuberculosis (streptomycin)

Masters of microbes

Task: To read and conduct research about some scientists and their contribution to medical history.

It is hard to believe, in our widely sanitised world, that there was a time when no-one knew that micro-organisms existed.

▶ **What to do:**

1. (a) Read each of the paragraphs below.

   (b) Write one sentence to explain each man's contribution to medical science.

   Anton van Leeuwenhoek (1632–1723) was a Dutch tradesman with no formal scientific education, yet he discovered the existence of bacteria and other micro-organisms. Although he did not invent the microscope, the ones he made could magnify over 200 times. He is known as 'the father of microbiology'.

   _____

   _____

   Edward Jenner (1749–1823) developed a vaccine against the deadly disease smallpox. He observed that dairymaids who were infected with a milder infection, cowpox, made a full recovery and were immune to smallpox.

   _____

   _____

   Louis Pasteur (1822–1895) discovered the 'germ theory' of disease; that the most infectious diseases are caused by germs. He also gave his name to pasteurisation, the heat process he discovered that destroys harmful micro-organisms in perishable foods, without destroying the foods themselves.

   _____

   _____

   Joseph Lister (1827–1912) became known as 'the father of antiseptic surgery' when he made the connection between the lack of hygiene in hospitals and the number of deaths following surgery—known as 'ward fever'. He insisted that surgeons washed their hands thoroughly before and after surgery and introduced the use of a fine carbolic acid spray during surgery.

   _____

   _____

2. (a) Research to find which famous scientist is credited with each medical breakthrough. Draw lines to match.

Alexander Fleming •	• transmission of yellow fever virus by mosquitoes
Louis Pasteur •	• discovery of polio vaccine
Walter Reed •	• discovery of penicillin
Jonas Salk •	• discovery of rabies vaccine
Selman Waksman •	• discovery of an antibiotic effective against tuberculosis

   (b) On separate paper, create a time line to show when these discoveries were made.

Tracing a food chain

Life and living • Life and living • Life and living • Life and living • Life and living • Life and living

Objective: Observes a local habitat to trace a food chain back to the producer.

Materials
- notebook and pen, habitat to observe, small trowel, small gardening fork

▼ Motivate
- Ask pupils to consider a local habitat: Is anything alive in this habitat? (plants, small creatures) What are the relationships between these things? How might the removal of one thing affect the survival of another?

▼ Experience
- Pupils create their initial labelled drawings of the habitat.
- Pupils record any immediately obvious creatures they see, then conduct a closer examination by carefully looking under foliage, stones, branches etc. Using the fork and trowel, he/she can also examine the soil.
- Close observation over a period of time, which may involve different times of day, will reveal evidence of a food chain such as plant → aphid → frog.
- Pupils draw their food chains using arrows to link one part of the chain to the next.

▼ Explain
- Pupils' observations will show feeding relationships between some of the creatures in their chosen habitat.

> A food chain represents the feeding order within an ecosystem.
>
> Plants are the organisms that are always at the beginning of any food chain as they produce their own food through photosynthesis. Such organisms are known as autotrophs.
>
> Plants are called 'producers' and all creatures, whatever their place in a chain, are called 'consumers'.
>
> Herbivorous animals eat the plants. Carnivorous animals eat the herbivores and other carnivores. Omnivorous animals eat animals and plants.
>
> Although there may be competition between animals for food sources, there is also an interdependence; if one species is removed, the whole chain may be affected.
>
> A food web is a network of interlinked food chains.

▼ Apply
- Pupils observe a different type of habitat to compare and contrast food chains.
- Pupils research on the Internet to compare and contrast a different type of habitat.

▼ Review and reflect
- Pupils complete a science recount on page xiii.
- Pupils reflect upon how they feel about the effect loss of habitat has on animal species; for example, when land is cleared for housing developments.

▼ Answers
What I've learned
1. beginning
2. food
3. producers
4. living
5. consumers
6. eat

Tracing a food chain

Task: To trace a food chain back to the producer.

You will need
notebook and pen
habitat to observe
small trowel
small gardening fork

▶ **What to do:**
1. Sketch and label a picture of the features in your habitat; e.g. plants, rocks, water.

2. Carefully examine and observe the habitat and record all the creatures you see there.

▶ **What happened?**
Draw a food chain you observed in your habitat.

▶ **What I've learned**
Use words from the box to fill the gaps.

| consumers | producers | beginning | eat | living | food |

Plants are always at the _____ [1] of a food chain. They do not eat other things, but produce their _____ [2] by photosynthesis. Plants are called _____ [3] because they produce food for other _____ [4] things to eat. These other creatures are called _____ [5] because they _____ [6] the plants and each other.

Prim-Ed Publishing ~ www.prim-ed.com   Hands-on science   **45**

Gardening gurus

Life and living • Life and living • Life and living • Life and living • Life and living • Life and living

Objective: Considers plant requirements when choosing the best location for planting.

Materials

- several garden centre plant labels detailing sunlight, shade and watering requirements; and expected growth (width and height). All labels must be washed clean of any soil or compost. Ensure there is a good mixture of labels for plants with differing sunlight needs.
- gardening reference books for further information on plants
- coloured pictures of the plants that correspond with the labels

▼ Motivate

- Ask: What do living things need to grow? (sunlight, air, water, nutrients)
- Ask: How have plants adapted to meet these basic needs? (leaves for breathing and feeding-photosynthesis, roots and stems to draw water and nutrients into the plant from the soil)
- Ask: In what ways have different plants adapted to their environment? (cactus – thin spiky leaves to reduce water loss in arid desert conditions; vines – climb close to the top of trees to access more sunlight)
- Ask pupils to look at the labels and discuss different ways of categorising the plants.

▼ Experience

- Pupils follow the worksheet to plan the layout of the flower garden and appreciate that different plants have different requirements.
- By looking at pictures and researching in the gardening reference books, pupils can suggest how the appearance of each plant may be related to its preferred growing conditions.
- Pupils will need to study the labels carefully to find out how much growing width each plant requires as this will determine how many plants pupils will require.

▼ Explain

- Pupils complete the 'What I've learned' section of the worksheet.

> An adaptation is something that helps a plant or animal survive and reproduce in its specific environment. Functional adaptations help the plant or animal survive in its environment; reproductive adaptations help it reproduce to propagate the species.
>
> Examples of functional adaptations are those for water-limited environments such as deserts, light-limited environments such as dense forest and woodland, supportive adaptations for plant root systems and defensive adaptations against predators.
>
> Examples of reproductive adaptations are methods for attracting insects such as scent, colour, size; strategies for preventing self-pollination.

▼ Apply

- Pupils choose one plant from each category (full sun, full shade, part sun/part shade) to research; sketch and label its parts; write a brief report on the life cycle of each plant; and draw up a graphic organiser to compare and contrast the three plants.

▼ Review and reflect

- Pupils complete a scientific diagram, on page xvi, to display and label the main parts of each of the three chosen plants.

Gardening gurus

Task: To plan the layout of a flower garden.

You will need
plant labels
gardening reference books

▶ **What to do:**

1. Study the plant labels and use the books for extra information on how to care for each plant.
2. Categorise plants based on sunlight requirements.
3. Give each plant a key symbol.

Full sun	Key	Part sun/Part shade	Key	Full shade	Key

4. Decide the best location for the plants on the flower garden plan below. (You do not need to use all of the plants but the plan must be covered.)
5. Determine how many of each plant will be required.
6. Use your key symbols to sketch each plant on the plan.

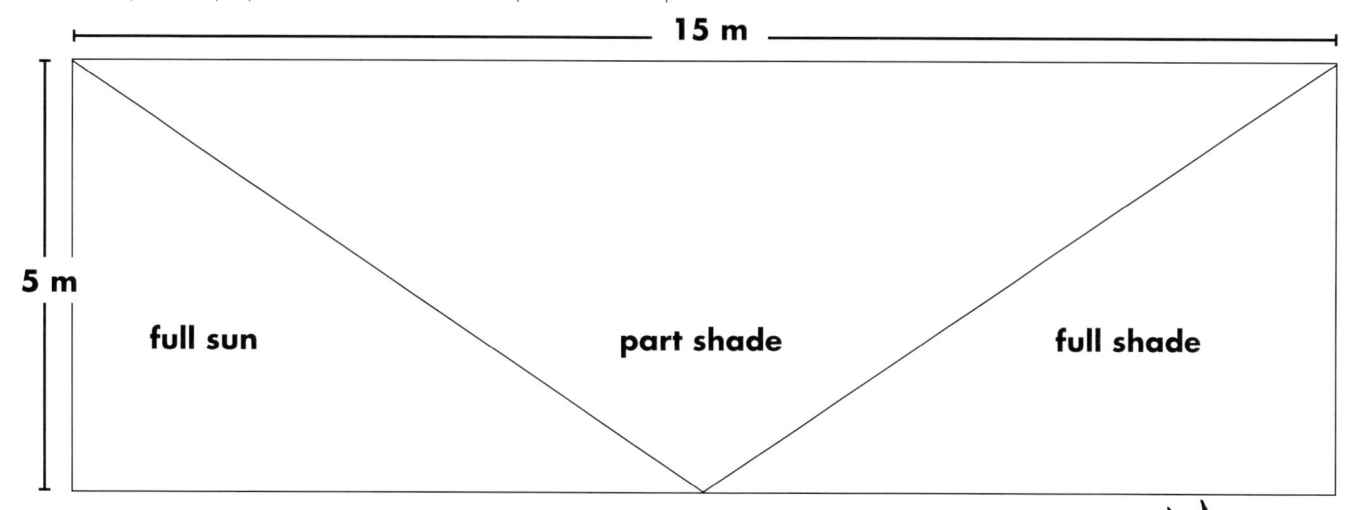

15 m

5 m

full sun   part shade   full shade

Points to consider:
- How tall will the plants grow?
- How far apart should they be planted?
- How much water do they need?

▶ **What I've learned**

What conditions are important to consider when planning a flower garden?

The year of a fruit tree

Life and living

Objective: Demonstrates stages associated with one year of a fruit tree's life by way of movement and dance.

Materials
- large working area, percussion instruments, music CDs, CD player, pencil and notepad, reference books and posters with illustrations depicting the various stages of life for a fruit tree.

▼ Motivate

- Ask pupils to express the feelings they associate with the appearance of trees at different stages of the year; e.g. hope and anticipation in the spring, the playfulness of autumn, bleakness in the winter.
- Ask: How do different animals (humans included) benefit from trees throughout the year? e.g. birds nesting and finding caterpillars in the spring and feeding from fruit in the summer; small mammals storing food in the autumn; humans harvesting the fruit in spring, enjoying the beauty and shades of the foliage while kicking through fallen leaves in autumn.
- Starting from the awakening of the tree in the spring, what stages does a tree go through in one year?

▼ Experience

- Allow groups time to plan and practise their performances. The amount of time and teacher input depends on their experience and confidence.
- By being physically involved in a performance, the pupils will be learning through the development of bodily-kinaesthetic intelligence.
- Pupils will use their knowledge of levels, weight and space in movement to illustrate the different stages.
- Pupils can choose to perform their own music, or use CDs or a combination of both. The tone and volume of music should reflect each movement.

▼ Explain

Different types of trees have different yearly cycles. Some trees are deciduous, meaning they loose their leaves during autumn and winter, while others are evergreen and do not shed their leaves according to the season. Many fruit trees (including pears, plums, apples, peaches and cherries) are deciduous. For these trees, the same annual process occurs, year after year. There are six main elements; from the appearance of buds, to the development of fruit and, finally, the fall of old leaves.

Bud break
During winter, most fruit trees remain dormant and have neither fruit nor leaves. When temperatures start to rise, during spring, both little fruit and leaf buds start to appear and open. These will form the new fruit and leaves for the tree.

Flower and leaf development
During spring, the immature flower develops from the bud and is closed and protected by the sepals; a green covering that surrounds the flower before it opens. The flower then blossoms, ready for pollination. Leaves also develop and grow.

Pollination
Blossoms produce a great deal of pollen. Pollen is required for flowers to become fertilised and for plants to reproduce. Some plants can self-pollinate, but most require cross-pollination. Once blossoms have become fertilised, the flowers fall away.

Fruit
In place of the flowers, fruit starts to grow. Fruit are the ovaries for seed and grow to allow the tree to reproduce. As fruit matures and ripens, it generally changes colour.

Seed drop
During late spring or summer, the seeds are shed by the plant, either by people or animals collecting the fruit and eating it or by the fruit ripening and dropping to the ground. The seeds are dispersed, allowing for new trees to form.

Foliage lost
As summer finishes and the weather cools, the leaves start to wither and change colour. Eventually, the leaves fall and the trees become dormant for winter, awaiting spring and for the process to start again.

▼ Apply

- Pupils plan a similar performance from the opening of the flower bud, pollination by insects, explosion of seeds and death of the flower. They direct pupils in a younger class to present this performance.

▼ Review and reflect

- Pupils complete a science reflection sheet on page xi.

The year of a fruit tree

Task: To create a music and dance performance which illustrates the different stages that occur in one year of a fruit tree's life.

You will need
large working area
percussion instruments
music CDs
CD player
pencil and notepad

▶ **What to do:**

1. In groups, discuss and practise the movement and music appropriate for each stage in the tree's cycle. Make brief notes as required.

2. In the table, describe your choice of movement and music for each stage. Include any props you will use to enhance the performance.

Stage	Movement	Music	Props
leaf buds appear and grow			
blossom buds appear and open			
insects pollinate the blossom and fertilisation occurs			
blossom dies and fruit develops			
fruit swells and ripens			
fruit eaten by birds and seeds dispersed			
fruit falls from trees and seeds dispersed			
fruit picked by humans and seeds dispersed			
leaves change colour			
leaves wither and die			
leaves fall from trees			

3. Present your performance to the rest of the class.

4. Rate your performance out of ten. _____/10

Healthy options

Life and living • Life and living • Life and living • Life and living • Life and living • Life and living •

Objective: Demonstrates understanding of a healthy, balanced diet by planning a menu for a weekend of meals.

Materials
- selection of healthy recipe books, supermarket brochures, healthy eating chart, 6 coloured pencils or highlighters, notebook and pen

▼ Motivate
- Show pupils photographs of food from recipe books and brochures. What feelings do they give rise to?
- Give 30 seconds for pupils to list as many unhealthy foods as possible. Repeat for healthy foods. Which list is longer? Do some pupils disagree on the categories of some foods? Can pupils explain reasons for their choices?
- Revise content of healthy food chart.

▼ Experience
- Pupils study brochures and recipe books to create menus for each meal time on separate paper.
- Having devised a colour key, pupils can highlight the ingredients in the table. This will make it easy for him/her to see how balanced and healthy their menus are.

▼ Explain
- Pupils should be able to describe the basis for a realistic healthy eating plan which allows for small, infrequent amounts of 'forbidden' foods.

A healthy, balanced diet should include:
- bread, cereals, pasta, rice and noodles; sources of carbohydrate required for energy
- vegetables and legumes; sources of fibre for healthy digestion
- fruit and vegetables; sources of essential vitamins and minerals for general health and wellbeing
- dairy produce; sources of calcium and protein, necessary for healthy teeth, bones and muscles
- meat, fish, poultry, eggs and nuts; sources of protein for health of muscles and blood.

Adequate amounts of water should be consumed to ensure full hydration. A small decrease in hydration level can result in a large drop in bodily function efficiency. By the time you feel thirsty, hydration levels have already dropped so far that the body is not functioning efficiently.

Although treats should be allowed, to keep the plan realistic, fats, oils and sugars should be used sparingly.

▼ Apply
- Plan a varied healthy lunch pack for each day of one week.

▼ Review and reflect
- Pupils complete a science reflection sheet on page xi.

Healthy options

Task: To prepare a healthy weekend menu for guests at a health retreat.

▶ What to do:

1. Read through the recipe books and brochures for ideas for each meal. Make notes and write a menu on a separate sheet of paper.
2. Use the table to list the main ingredients for each meal.

Friday	dinner	
Saturday	breakfast	
	lunch	
	dinner	
Sunday	breakfast	
	lunch	

You will need

- selection of healthy recipe books
- supermarket brochures
- healthy eating chart
- notebook and pen
- 6 coloured pencils or highlighters

3. Devise a colour key for each food group and highlight the ingredients in the table with the appropriate colour from your key.

☐ bread, cereals, rice, pasta, noodles ☐ vegetables, legumes ☐ fruit

☐ dairy produce ☐ meat, fish, eggs, nuts ☐ fats, oils, sugars

▶ What happened?

How healthy is your weekend menu planner?

Check your plan against the healthy eating chart and give it a score out of ten. ____/10

▶ What I've learned

What is a healthy eating plan?

Upside down views

Life and living • Life and living • Life and living • Life and living • Life and living • Life and living •

Objective: Creates a pinhole camera to illustrate the inverted image received by the retina.

Materials

- 2 cardboard cylinders – one of longer length but smaller diameter which can slide inside the other cylinder; scissors, 25-cm² piece of baking paper; 25-cm² piece of aluminium foil; 5 cm x 25 cm rectangle of black construction paper fanned into 1 cm folds; two elastic bands; thin needle; sticky tape
- diagram of the eye displaying the cornea, iris, pupil, lens, retina and optic nerve

▼ Motivate

- Ask: What would life be like if we were all upside down? How would we move around, ride a bicycle, eat our food?
- Ask: How confusing would it be if we were the right way up but saw everything else as upside down? How would we know how to step on a bus, eat a meal at a table or jump in a pool? But this is what we actually see through our eyes! So why aren't we confused?

▼ Experience

- Pupils follow the procedure to make the pinhole camera, which gives an inverted image of an object.
- The size of the pinhole will determine the clarity of the image. If it is too large, too much light will penetrate and the image will be blurred or even indistinguishable. If the pinhole is too small, insufficient light will penetrate to illuminate the screen.
- The black construction paper around the viewing end of the outer cylinder will help to prevent further light from entering the camera, resulting in a clearer image.

▼ Explain

- Light rays from the top of the object pass through the pinhole to the bottom of the cylinder's screen. Light rays from the bottom of the object pass through the pinhole to the top of the screen, thus inverting the image.
- This activity demonstrates how the human eye works.

> The cornea is a thin, protective layer on the front of the eye. Light enters the eye through the pupil, a hole in the iris which controls the amount of light entering the eye by altering its size.
>
> Light rays pass through the lens of the eye, where they are bent to form an inverted image on the retina. The lens can change shape to focus on objects at various distances.

> Although the image received on the retina is inverted, the brain immediately inverts it so we see everything the right way up.

▼ Apply

- Use pins of different thickness to determine which hole size produces the clearest image.

▼ Review and reflect

- Pupils complete a scientific diagram, on page xvi, to illustrate their understanding of how the pinhole camera and human eye work.

▼ Answers

What happened?

1. Images are inverted
2. The image changes size.

Upside down views

Task: To make a pinhole camera.

▶ What to do:

1. Follow the instructions to make the pinhole camera.

(a)	Use elastic bands to secure the baking paper over one end of the cylinder with the smaller diameter and secure the aluminium foil over one end of the wider cylinder.	
(b)	Carefully use the needle to make a small pinhole in the centre of the aluminium foil cover.	
(c)	Slide the smaller cylinder inside the larger one.	
(d)	Secure the construction paper around the open viewing end of the cylinder with strips of sticky tape.	

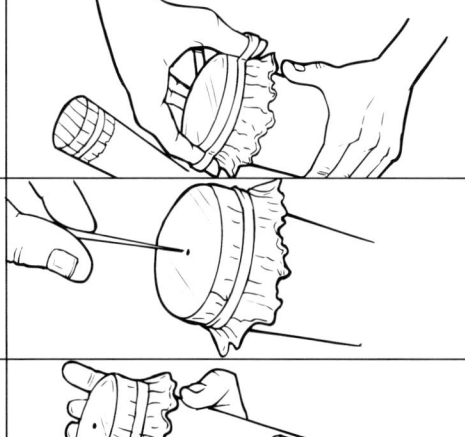

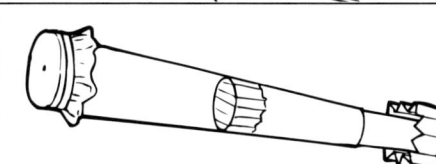

You will need

2 cardboard cylinders – one of longer length but smaller diameter which can slide inside the other cylinder

scissors

25-cm² piece of baking paper

25-cm² piece of aluminium foil

5 cm x 25 cm rectangle of black construction paper fanned into 1-cm folds

2 elastic bands

thin needle

sticky tape

2. On a sunny day, view objects through your camera.

▶ What happened?

1. What do you notice about the images you see through the camera?

   _____

   _____

2. How do the images change when the inner cylinder slides back and forth?

   _____

   _____

▶ What I've learned

Draw a labelled diagram explaining how the pinhole camera works.

Prim-Ed Publishing ~ www.prim-ed.com   Hands-on science

Bony hands

Life and living • Life and living • Life and living • Life and living • Life and living • Life and living

Objectives:
- Learns the names of the bones found in the hand.
- Learns the location of the bones found in the hand.

Materials
- sharp pencil, scientific diagram of the skeleton of a hand

▼ Motivate
- Ask pupils to look at their hands. Can they distinguish and count the bones? Can they see/feel any pattern in the arrangement of the bones?
- Ask pupils to look at their feet. Repeat the questions.
- What similarities and differences can they see between the bones of their hands and their feet?

▼ Experience
- Pupils follow the directions on the worksheet to draw bones inside the hand outline.
- Show pupils a scientific diagram of the skeleton of a hand so they can compare and rate their drawings.
- Using their keys, pupils colour and label each type of bone.

▼ Explain

The adult human skeleton has 206 bones. Just over half of these bones are located in the hands and feet.

Each hand has 27 bones and each foot has 26 bones. The skeletal configuration of the hands and feet are very similar.

The two largest bones in the feet, the calcaneus and the talus, carry most of the body's weight.

The bones of the feet are held in position by ligaments, which bind the bones together, and by tendons at the ends of the muscles.

Throughout life, the human skeleton changes composition. At birth, a baby has 270 bones. However, as the child develops, at various sites in the skeleton some bones fuse together to become one.

Children's bones are softer than adult bones. As children grow, minerals are deposited and their bones become harder, from the centre out. This process is called ossification.

▼ Apply
- Pupils research the bones of the feet. Repeat the worksheet activity with the bones of a foot instead of a hand.

▼ Review and reflect
- Pupils complete a scientific diagram, on page xvi, to illustrate their understanding of how the bones of the hand are arranged.

▼ Answers

What I've learned

There are 54 bones in a pair of adult hands.

Bony hands

Task: Draw the skeleton of a hand by following a description of the hand's bones.

You will need

sharp pencil

picture of the skeleton of a hand to compare

▶ **What to do:**

1. Look at your own hands as you read the description of each type of bone in the hand.

   *Each finger has three long bones, but the thumb has only two. These bones are called **phalanges**.*

   *The palm of the hand has five long bones called **metacarpals**. The knuckles are the ends of the metacarpals.*

   *The wrist has eight small, irregular-shaped bones; an outer curve of five bones and an inner layer of three bones. These bones are called **carpals**.*

2. Use this information and observation to draw, in pencil, each bone in the outline of the hand below.

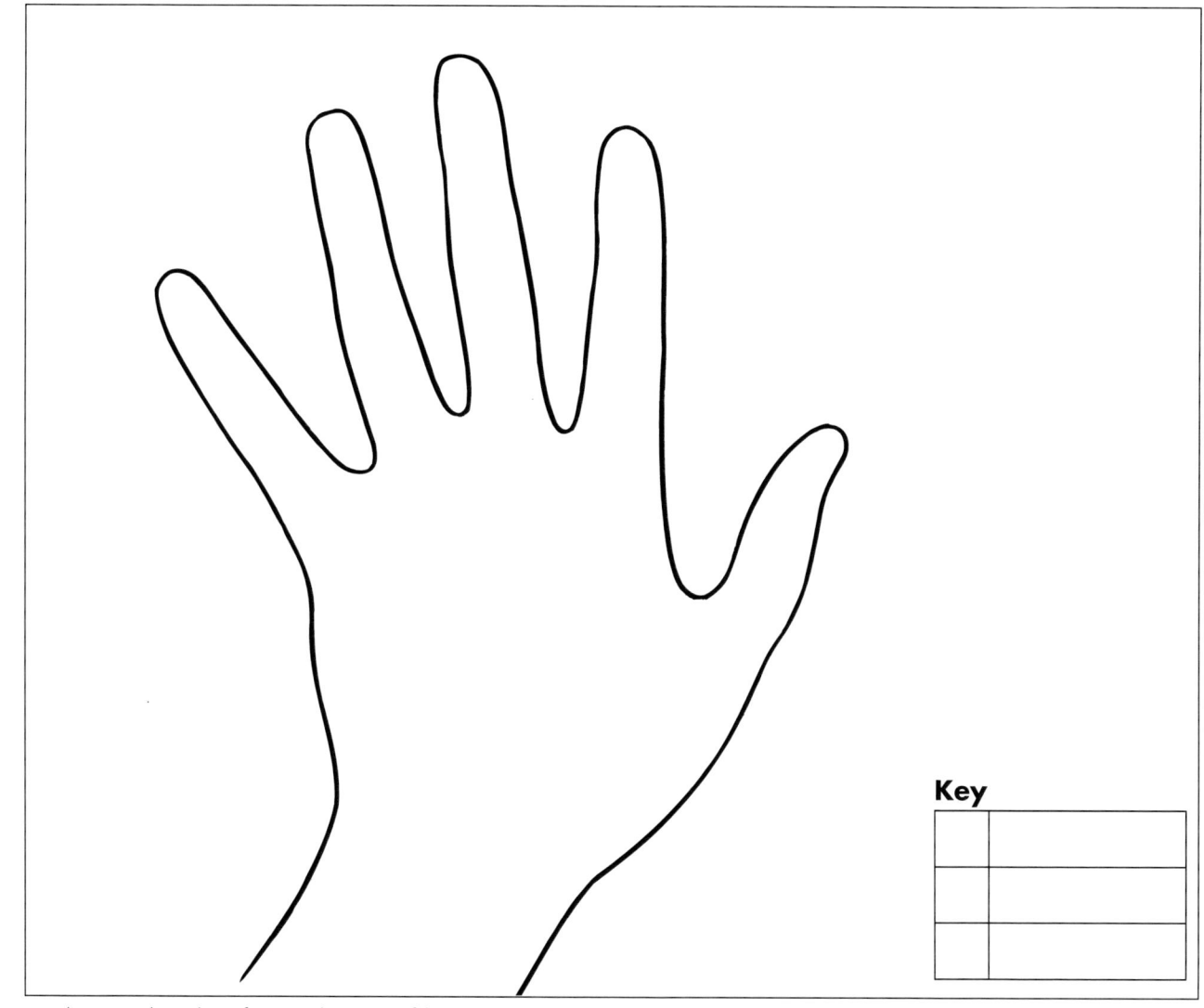

Key

3. Make a colour key for each type of bone and use it to colour the skeleton.
4. Label the groups of bones.
5. Compare your diagram with the teacher's picture. Give your diagram a mark out of ten. _____/10

▶ **What I've learned**

How many bones are there in a pair of adult hands? _____

Prim-Ed Publishing ~ www.prim-ed.com   Hands-on science

The money bridge

Natural and processed materials • Natural and processed materials • Natural and processed materials

Objective: Investigates how the properties of a material can be changed by altering its shape.

Materials

- plain paper (such as photocopier paper), 5 paperclips, 2 thick books, a collection of coins of the same value (or metal washers), ruler, scissors

▼ Motivate

- Display the single piece of paper and ask the class how many coins they think a bridge made from one piece of paper will hold. Write their estimates on the board.
- Demonstrate this by placing the piece of paper flat across two books placed about 20 cm apart. Add the coins, one-by-one, in the middle of the paper, counting each coin with the class. Continue until the bridge falls.

▼ Experience

- Write the number of coins the bridge held on the board. Organise the pupils into small groups and ask them to design a bridge that will hold the greatest number of coins.
- Pupils can:
  – use one sheet of paper only
  – use scissors
  – use none, some or all of the paperclips provided
  – fold, roll, cut or weave their paper

  Note: The paper must just rest on the books. It cannot be taped down. No tape can be used in this experiment.

▼ Explain

- A bridge must support its own weight (the dead load) as well as the weight of anything placed on it, such as the coins (the live load).
- To make their bridges stronger the pupils could:
  – 'accordion pleat' the paper (corrugating) and place the coins in the slots
  – roll the paper around the coins and fasten the ends with the paperclips.

> All materials have certain properties which make them useful. The materials used to build bridges need to be sturdy (strong) and also rigid (stiff).
>
> Although one single sheet of paper is very flexible and also very thin and weak, its shape can be altered. This changes the properties of the paper.
>
> There are a number of ways to make paper stronger, such as rolling, folding, corrugating, angling and creating shapes. Changing the paper's shape can also change the way the paper resists forces.

▼ Apply

- Pupils build bridges using other materials, such as straws and pins, and test how much weight they can carry.
- Pupils investigate the strength of different shapes. Survey the school building to see where different shapes have been used to provide strength. (The triangle is the most rigid shape.)
- Use the science investigation framework on page xiv to find out if bridges hold more weight when the weight is distributed along the length of the bridge or kept at a single point.

▼ Review and reflect

- Pupils write a recount of 'The money bridge' activity using the recount framework on page xiii.

The money bridge

Task: To build a bridge using a single sheet of paper that can hold the maximum amount of coins.

▶ What to do:

1. Place the books 20 centimetres apart.
2. Discuss with your group how you think your bridge should be constructed to hold the most coins. A maximum of 5 paperclips can also be used, if required, in your design.
3. How many coins do you believe your bridge will hold? ☐
4. Try different designs and choose your best one.
   Note: The paper must rest on the books, it cannot be taped down. (No tape can be used in this experiment.)
5. Test you bridge by loading it with coins until it collapses.

You will need

sheet of plain paper

5 paperclips

2 thick books

coins (or washers)

ruler

scissors

▶ What happened?

1. Draw and label your bridge. Use a ruler for all straight lines.

2. Explain your design. _____
   _____

3. (a) Our bridge supported ☐ coins.
   (b) The maximum number of coins supported in the class was ☐ coins, using a design where the paper was …
   _____

▶ What I've learned

1. Was your bridge as strong as you first believed?   **yes**   **no**
2. Why did it fail/succeed? _____
   _____
3. On the back of this sheet, redesign your bridge and test it with a new sheet of paper.

Crushing columns!

Natural and processed materials • Natural and processed materials • Natural and processed materials

Objective: Creates and tests three different column shapes to see which holds the most weight.

Materials

- For each pair: card (13-cm x 8-cm – index cards will work well for this experiment or any card of that thickness), ruler, jug, masking tape, sand, empty tin, weighing scales

▼ Motivate

- Demonstrate to pupils paper strength by continually folding a piece of computer paper until it cannot be folded any more. As the fibres in the paper are piled on top of each other, the paper appears stronger and thicker and cannot be folded.

▼ Experience

- Organise pupils into pairs and read the worksheet together as a class.

  Note: The 'What will happen' section must be completed before the experiment is conducted.

- Pairs follow the steps on the sheet to conduct and report on their investigation.

▼ Explain

Some materials are stronger than others, but if they are manipulated into certain shapes, the strength of that material can be changed. Columns are often found in buildings. They are used to support ceilings and roofs, and are usually made of concrete or wood.

Generally, the triangle is one of the most rigid and stable shapes used in construction. Rectangles can be strengthened by adding supports that form triangles within the rectangle, usually diagonally across the shape.

▼ Apply

- Give small groups a sufficient supply of paper. Ask them to build a bridge between two desks 50 cm apart. Groups can try: rolling, folding, corrugating, angling or creating shapes with their paper to build the strongest and most rigid bridge.

  Test what the bridge will hold. Compare bridges and discuss reasons for weak and strong bridges.

▼ Review and reflect

- Pupils complete the recount framework on page xiii about the 'Crushing columns!' activity.

Crushing columns!

Task: To test the strength of columns of various shapes.

Columns are often found in buildings. They are used to support ceilings and roofs, and are usually made of concrete or wood.

You will need

3 pieces of card (13 cm x 8 cm)

ruler

jug

masking tape

sand

empty tin (such as coffee tin)

weighing scales

▶ What will happen?

1. Which column shape do you think will support the most weight?

   | cylinder | rectangular | triangular |

2. Why? _____

   _____

▶ What to do:

1. Construct three 13-cm tall columns using the card and tape. Make:
   - one column a cylinder
   - one column a rectangular prism
   - one column a triangular prism

   Note: The prisms are open prisms (no ends) and a **maximum of 15 cm** of masking tape can be used to construct each column. (Measure it!)

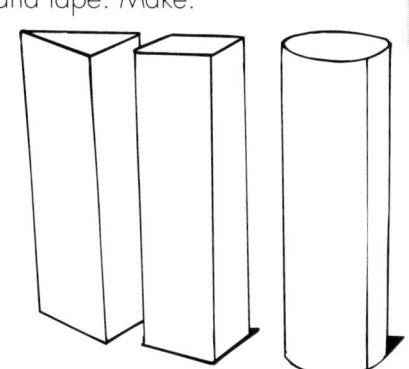

2. Place the tin on top of each column and use the jug to slowly pour sand into the tin until the column collapses.

3. Scoop any spilled sand back into the tin. Weigh the tin and record the results in the table below.

▶ What happened?

Column shape	Cylinder	Rectangular	Triangular
Weight held			

1. The shape that held the most weight was _____.

2. My prediction was | correct | | incorrect |

3. Class results

   The shape that held the most weight: ☐ cylinder ☐ rectangular ☐ triangular

▶ What I've learned

Write a short paragraph that summarises the class results and what you have learned about the strength of different shapes.

_____

_____

_____

Walking on eggshells

Natural and processed materials • Natural and processed materials • Natural and processed materials

Objective: Designs and performs a test to demonstrate the hardness of eggshells.

Materials

- For each group: lots of raw eggs, long needle, bowl, newspaper
- Other materials requested by the pupils once they have designed their experiments

▼ Motivate

- Ask pupils what the saying 'walking on eggshells' refers to? Does this mean that eggshells are very brittle and soft? Are they? Discuss.

▼ Experience

- Pupils work in small groups to create an investigation exploring the hardness properties of eggshells.

Note: This activity requires two lessons as it is advisable that the yolks are removed the day before and the eggshells thoroughly washed and dried overnight before pupils begin experimenting with them. Alternatively, ask parents to save their broken eggshells/halves and pupils can bring them to class. The shells do NOT have to be whole to conduct this investigation.

How to remove an egg yolk:

1. Firstly, wash the egg in soapy water and then dry it.
2. Insert a long needle in the wider end of the egg and make a small hole.
3. Make a slightly larger hole in the narrower end of the egg. Push the needle into the centre and move it around to break the yolk.
4. Hold the egg over a bowl with the narrower end facing down. Place your lips over the hole at the large end of the egg and blow firmly until all the egg comes out the hole at the small end. Hold the egg carefully!
5. Rinse out the egg by running a thin stream of water into the larger hole. Blow out the water. Dry the eggshell by propping it up in a dish drainer with the wider end facing down.

Safety note: NEVER ingest raw egg. Raw eggs can carry salmonella bacteria which can cause serious illness.

- For pupils having difficulty designing a test:
Try breaking the eggshells in half, leaving the dome-shaped ends. Place them on a flat bench. What will they carry? Try a thin book, then continue to increase the weight. How much will the eggs carry before cracks start to appear?

▼ Explain

- Materials have certain properties which make them useful. We use certain materials for certain jobs; for example: strong materials can be used to build bridges, flexible materials can be used to make fishing rods and belts and waterproof materials can be used to make umbrellas and tents.

Although eggshells aren't very thick, they are actually quite hard.

The standard bird eggshell is approximately 95% calcium carbonate crystals, which gives it strength and stiffness but also leaves it brittle.

If arranged correctly, eggshells can carry a remarkably heavy load. The arch or dome-shaped ends of the eggshell give it great strength.

▼ Apply

- Pupils investigate and compare the hardness of different types of eggshells, such as white, brown, free-range, battery-produced, duck etc.

▼ Review and reflect

- Pupils present an information poster displaying knowledge, data, summary and conclusion of the investigation.
- Images can also be included in the poster; use a digital camera to photograph the pupils conducting their investigations.

Walking on eggshells

Task: To design a test to determine the hardness of eggshells.

Group members

Materials

What we already know about eggs and eggshells

What we will do

Diagram of test (with labels)

What happened?

What would you do differently next time?

How hard are eggshells?

weak   quite hard   super hard!

Matter mosaics

Natural and processed materials • Natural and processed materials • Natural and processed materials

Objectives:
- Understands that matter can exist as a solid, liquid and gas.
- Demonstrates knowledge of the structure of the three states of matter by creating a mosaic.

Materials
- coloured paper, hole punch, scissors, glue

▼ Motivate
- Divide the board into three columns with the headings 'solids', 'liquids' and 'gases'. Ask pupils to volunteer to draw and label a picture in one of the columns. When all columns have a number of pictures in them, ask the class to describe the differences between the three states of matter.

▼ Experience
- Read the explanatory text at the top of the worksheet with the class.
- Pupils use a hole punch to make 'molecules'.
- Pupils glue the circles to the page to show the behaviour of solid, liquid and gas molecules.

▼ Explain

> Everything (all matter) is made up of tiny particles called molecules. Molecules are always moving.
>
> In a solid, molecules are tightly packed together in a definite shape and are usually easy to handle. Solids (like pencils, desks and cars) are rigid and hold their shape.
>
> Liquid molecules are close together, but can slide past each other and change their location. This means liquids can flow and change shape easily. They take on the shape of their container.
>
> Gas molecules are very widely-spaced. Gases are often difficult to sense or see, but we know they are there. They are usually detected through our sense of smell. Gas molecules spread out in all directions, but can fit and even be squashed to fit into different shapes.

▼ Apply
- Pupils draw a picture of a fish tank (with water, fish, weeds and air bubbles). Pupils make a mosaic using one colour of paper-punch dots for the solids, another for the liquids and another for the gases.

▼ Review and reflect
- Outside, draw a large circle on the ground in chalk. Ask pupils to stand inside the circle and behave as a:

  solid (should be standing very close together to fill the circle and vibrating)

  liquid (should be less pupils; some touching but moving around the circle sliding past each other)

  gas (less pupils again; each moving rapidly with lots of room to move around)

Matter mosaics

Task: To display knowledge of the structure of solids, liquids and gases by making matter mosaics.

Everything (all matter) is made up of tiny particles called molecules. Molecules are always moving. In a solid, molecules are in a definite shape and are so tightly packed together they can only move by vibrating against each other. In a liquid, molecules are close together and touching, but can slide past each other allowing liquids to flow and change shape easily. In a gas, molecules are widely-spaced, meaning a gas can spread out quickly (diffuse) and even be squashed.

What to do:
Use coloured paper circles to turn these examples of a solid, liquid and gas into colourful 'matter mosaics'.

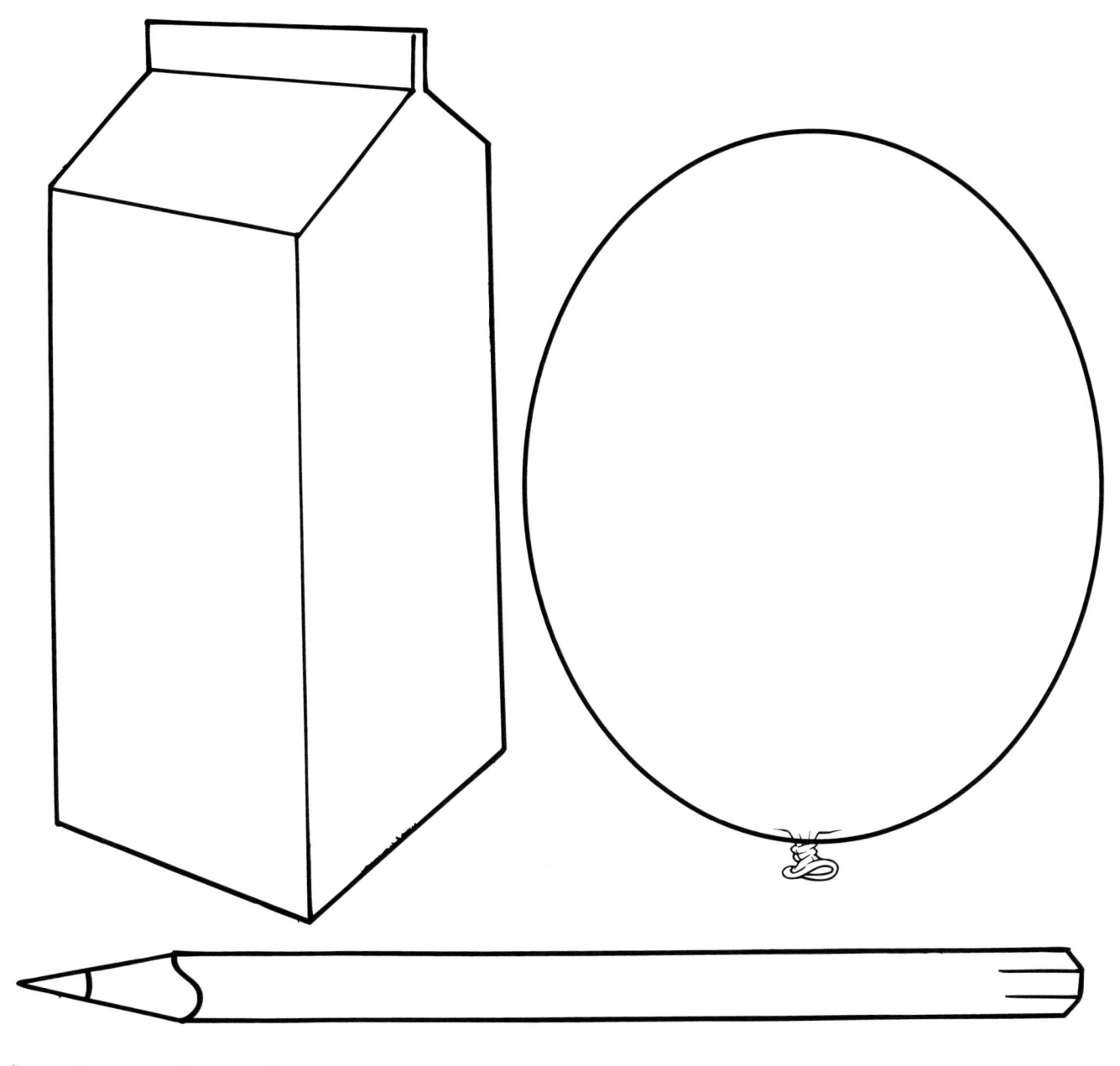

What I've learned
1. The molecules in a _____ can vibrate and take the shape of a container.
2. The molecules in a _____ are far apart and can move freely.
3. The molecules in a _____ cannot move away and have a fixed shape.

Prim-Ed Publishing ~ www.prim-ed.com   Hands-on science

Dissolving rock

Natural and processed materials • Natural and processed materials • Natural and processed materials

Objective: Conducts an experiment to test if a liquid can dissolve a rock (chalk).

Materials
- For motivation: hard-candy sweets, zip lock bags, heavy object (like a rolling pin)
- For worksheet: 3 pieces of long, white chalk; vinegar; lemon juice; 3 identical glasses (labelled), water

▼ Motivate
- Give each pair of pupils two hard-candy sweets (such as humbugs). Place one in a zip lock bag and crush it by pressing on it with a heavy object. At the same time, one pupil puts the whole sweet in their mouth and the other pupil puts the crushed sweet in their mouth. Without chewing, which pupils' sweet dissolves first?

Safety note: Be aware of any pupils who are unable to eat sugar.

▼ Experience
- Pupils make their predictions on the worksheet and then follow the instructions to complete the experiment.

▼ Explain
- Provide pupils with the following explanation before they complete the 'What I've learned' section of the sheet.

> Chalk is made from the rock limestone (or calcium carbonate). When an acid (vinegar or lemon juice) is combined with the limestone, a chemical reaction occurs and the gas carbon dioxide is formed. The reaction can make the rock become soft, weakening it and eventually breaking it apart so it crumbles—a process called erosion.

▼ Apply
- Pupils design a test to discover if other rocks contain carbonates. Pupils conduct their test on rocks found around the school. They record their data and present their findings to the class.

▼ Review and reflect
- Pupils complete a science reflection on page xi about the 'Dissolving rock' activity.

▼ Answers
What I've learned
1. Teacher check.
2. Lemon juice and **vinegar** are both acids. Chalk is made from the rock **limestone** (or calcium carbonate). When an **acid** is combined with the limestone (a carbonate), the gas carbon dioxide is **formed**. This reaction causes the rock to **weaken** and eventually break apart and **crumble**. This is called **erosion**.

Dissolving rock

Task: To test if liquids can dissolve rock.

▶ What to do:

1. Label one glass 'water', another 'lemon juice' and the third 'vinegar'.
2. Pour the corresponding liquid into each glass.
3. Place one piece of chalk in each glass. The chalk should be long enough so part of it is not in the liquid.
4. Place the experiment where it will not be disturbed and observe it each day for four days.

You will need

3 pieces of long, white chalk

vinegar

lemon juice

3 identical glasses (labelled)

water

▶ What will happen?

What do you think will happen to each piece of chalk? Include a reason for your prediction.

Chalk in water	Chalk in lemon juice	Chalk in vinegar

▶ What happened?

Describe the chalk on the final day of your observations.

Chalk in water	Chalk in lemon juice	Chalk in vinegar

▶ What I've learned

1. Can a liquid dissolve rock? **yes** **no**

2. Use the following words to complete the paragraph.

   | formed | weaken | vinegar | erosion | limestone | acid | crumble |

   Lemon juice and _____ are both acids. Chalk is made from the rock _____ (or calcium carbonate). When an _____ is combined with the limestone (a carbonate), the gas carbon dioxide is _____. This reaction causes the rock to _____ and eventually break apart and _____. This is called _____.

Separating colours

Natural and processed materials • Natural and processed materials • Natural and processed materials

Objective: Conducts experiments to explore the many colours that are used to create inks and dyes.

Materials
Splitting inks: coloured felt-tip pens, blotting paper, saucer, water
Smarties™ split: 1 Smarties™, coffee filter paper, water, plate
Ready, set ... race! 3 different types of black marker pen, coffee filter paper/blotting paper, 2 pegs, glass, pencil

▼ Motivate

- Ask the class:

  Is the advertising phrase that 'M&M's™ only melt in your mouth and not in your hand' true?

  How do you know? What have you noticed about the colours left in your palm? Do they always match the M&M™ you were holding? Experiment by holding an M&M™ in a warm palm.

▼ Experience

- Organise equipment into separate trays for each experiment and label them.
- Organise pupils into groups.
- Pupils choose one experiment and follow the steps, recording their observations on the sheet.
- When completed, groups swap their equipment and conduct another experiment until all three are completed.

▼ Explain

> When certain solids are mixed together, they can sometimes be separated by using a process called chromatography. This involves the substance being dissolved in a liquid (a solvent) which then carries the solid along a strip of blotting or filter paper. The solids which dissolve in the liquid the best, will be carried along the paper the furthest. Colours can be separated to show the different components that have been combined to make an ink, dye or paint.

▼ Apply

- Pupils create an investigation to discover how food colouring or water-based paints can be separated into their colour components. Use the science investigation framework on page xiv. Pupils conduct the experiment and report their findings to the class.

▼ Review and reflect

- Pupils choose their most successful experiment and conduct it with a small group of younger pupils. Pupils explain the results to the group using simplified terms.

▼ Answers

Splitting inks

- The colours will move up the paper and away from the outline of the butterfly. The colours will also separate into different colours that have been combined to make that particular colour pen.

Smarties™ split

- Rings of different colours will form on the paper around the Smarties™.

Ready, set ... race!

- The inks will move up the paper and separate into different colours.

Separating colours

Task: To conduct experiments that separate inks and dyes into their colour components.

▶ **What to do:**

Choose a 'separating colours' experiment and follow the procedure.

Splitting inks

You will need:

marker pens   blotting paper
saucer   water

What to do:

1. Draw a butterfly on blotting paper and colour it in using marker pens.
2. Pour a little water into the saucer.
3. Dip just the bottom of the paper in the water and hold it there for a few minutes.

What happened?

_____
_____
_____

Smarties™ split

You will need:

1 Smarties™   coffee filter paper
water   plate

What to do:

1. Place the filter paper on the plate.
2. Place a Smarties™ in the centre.
3. Dip a finger into the water and hold it above the Smarties™ so water drips onto it.
4. Keep dripping water on the Smarties™ until water begins to spread out on the sheet—then wait!

What happened?

_____
_____
_____

Ready, set ... race!

You will need:

3 different black marker pens   glass   pencil
coffee filter paper or blotting paper   pegs

What to do:

1. Cut the paper into a rectangle so it fits in the glass with a section above the top of the glass.
2. With the pencil, draw a line about 3 cm from the bottom of the paper.
3. Draw a dot with each pen on the line, each a good space apart from the other.
4. Clip pegs to each side of the top of the paper and rest it in the glass (with the paper just touching the bottom). Keep adjusting the pegs until this happens.
5. Remove the paper and add about half a centimetre of water (0.5 cm) to the glass.
6. Return the paper to the glass. The bottom should rest in the water (but not the dots!) Wait!

What happened? *(And which pen won the race?)*

Glue test paper here

_____
_____
_____
_____
_____
_____
_____
_____

Making a gas

Natural and processed materials • Natural and processed materials • Natural and processed materials

Objective: Combines acids and carbonates to create the gas carbon dioxide.

Materials
- vinegar, lemon juice, cola-flavoured soft drink, grapefruit juice, baking soda, bicarbonate of soda, washing powder, crushed eggshells, glass jars and jugs, labels, pen

▼ Motivate
- Open a bottle of fizzy soft drink in front of the class. Ask the pupils to describe what they see and hear.
- Ask: What happens after a number of days? (The drink goes 'flat'.) What does 'flat' mean? Which gas makes the soft drink 'fizzy'? (carbon dioxide)

Note: Some prior knowledge of acids would be beneficial before the activity.

▼ Experience
- Set up the acids and carbonates in glass jugs and jars. Label each substance clearly.

> **Safety note:** Make pupils aware that washing powder can burn and should never be tasted. Hands must be washed thoroughly after the experiment.

- Discuss how pupils will know if a gas has been created. (Pupils will see bubbles and hear a fizzing sound.)
- In groups, pupils use three different combinations of materials to make a gas.

▼ Explain
- Provide pupils with the following explanation before they complete the 'What I've learned' section of the sheet.

> When some materials are mixed together a gas is made. In these chemical reactions, an acid and a carbonate react together and produce the gas carbon dioxide.

▼ Apply
- Pupils make and test sherbet. How do they know a chemical reaction has occurred and a gas has been created?

Sherbet recipe (per pupil)

$1/4$ teaspoon bicarbonate soda

$1/2$ teaspoon citric acid powder

3 teaspoons of icing sugar

Mix all three ingredients together thoroughly.

Note: strict hygiene must be observed as the experiment will be tasted.

> When water (saliva) is added to the mixture, a fizz and bubbles appear showing that a chemical reaction has occurred.

▼ Review and reflect
- Complete the report framework on page xii for the sherbet activity.

Making a gas

Task: To create the gas carbon dioxide.

▶ What to do:

1. Choose any liquid from the 'Acids' box and mix it with something from the 'Carbonates' box to make the gas carbon dioxide. Record what you did on the table below.

Acids	
vinegar	lemon juice
cola-flavoured soft drink	grapefruit juice

Carbonates	
baking soda	bicarbonate of soda
washing powder	crushed eggshells

2. Repeat Step 1 two more times and complete the table.

	Acid	Carbonate	What happened?	Gas created?
1				yes / no
2				yes / no
3				yes / no

▶ What happened?

The acid and carbonate mix that:

- made the loudest fizzing sound was _____ + _____

- created the most bubbles was _____ + _____

- produced the most carbon dioxide gas was _____ + _____

- produced the least carbon dioxide gas was _____ + _____

▶ What I've learned

When an acid is mixed with a carbonate _____

_____

Prim-Ed Publishing ~ www.prim-ed.com · Hands-on science

Rotten rust

Natural and processed materials • Natural and processed materials • Natural and processed materials

Objectives:
- Conducts an experiment to produce an irreversible chemical change—rust.
- Understands that air and water must both be present to produce rust.

Materials
- 2 jars, 2 nails, water, kettle, vegetable oil
- Note: This experiment needs to be completed over one week.

▼ Motivate
- Ask pupils what happens to their bicycles if they are left out in the weather over time? (They rust.)
- Where do they rust? Where don't they rust? Why? (Painting a metal such as iron and steel can stop the corrosion process.)

▼ Experience
- Pupils follow the steps on the worksheet to conduct the experiment.

  Note: When water is boiled, the dissolved oxygen is removed. Pupils should pour the boiled water into Jar B carefully and slowly to make sure they do not reoxygenate it.

- Once the experiment has been conducted, as a class, discuss the conditions of each jar. The nail in Jar A is in tap water with no covering (nail has moisture and air); whereas, the nail in Jar B is in water which has been boiled and has a layer of oil on top (just moisture, no air).

  Safety note: Pupils should always take care when using hot water.

▼ Explain
- Pupils complete the 'What I've learned' section of the worksheet.

> Rusting is a chemical reaction. Chemical reactions (or changes) are an irreversible change meaning a new substance has been produced—rust!
>
> When iron is left in the air and moisture, a flaky, reddish-brown substance forms on its surface called rust. Rust is caused by corrosion—a slow chemical change in which iron atoms combine with oxygen atoms from the air—forming iron oxide or rust. If left untreated, corrosion can weaken the strength of the iron.

> Steel is mainly made of iron and so is prone to rusting.
>
> A paint covering will stop the air from reacting with the iron and steel, and stop corrosion from occurring. Things made from iron are often painted to stop corroding.

▼ Apply
- Does salt water effect iron and steel in a similar way? Pupils use the investigation framework on page xiv to design an experiment to explore this idea. Pupils work in groups to conduct the experiment and report back to the class with their findings.

▼ Review and reflect
- Complete the science reflection sheet on page xi about the 'Rotten rust' activity.

▼ Answers
- The nail with moisture and air (nail in Jar A) will corrode faster than the nail in boiled water with the oil covering (nail in Jar B).

Rotten rust

Task: To conduct an experiment that produces an irreversible change to a material—rust!

You will need
2 jars
2 nails
water
kettle
vegetable oil

▶ What to do:

1. Label the jars, Jar A and Jar B. Place a nail in each jar.
2. Half fill Jar A with water from the tap.
3. Boil water in the kettle. Allow to cool.
4. Pour the same amount of boiled water into Jar B. Pour the water slowly, so no air bubbles appear.
5. Carefully pour a layer of vegetable oil on top of the boiled water in Jar B.
6. Observe the nails each day for one week. Record your observations below.

▶ What will happen?

What do you predict will happen to each nail? Give reasons for your answer.

_____
_____
_____

▶ What happened?

Record your observations and ideas below.

	Jar A	Jar B
Day 1 Date:		
Day 2 Date:		
Day 3 Date:		
Day 4 Date:		

▶ What I've learned

1. Which conditions resulted in the least amount of rusting?

   ☐ Jar A—moisture and air   ☐ Jar B—just moisture

2. What does this tell you about how rust is formed?

   _____
   _____
   _____

Ice-cream insulator

Natural and processed materials • Natural and processed materials • Natural and processed materials

Objective: Follows a recipe to create a baked ice-cream dessert with a meringue 'insulator'.

Materials

- ice-cream and scoop, round biscuits, 3 eggs, caster sugar, salt, cream of tartar, baking paper, biscuit tray, measuring spoons, cup measure, bowl, whisk, oven and oven gloves, spatula

▼ Motivate

- Ask the class if anyone has ever ordered 'baked ice-cream' in a restaurant. What does it look like? What does it taste like? Often this dessert is on the menu in Asian restaurants.
- Why doesn't the ice-cream melt in the hot oven?

▼ Experience

- Pupils follow the procedure to make baked ice-cream.

  Note: – To create the meringue, ensure eggs are at room temperature.
   – If the meringues start to become brown when in the oven, remove them immediately.

  Safety note: All utensils, benches and hands must be washed thoroughly and strict hygiene observed as this experiment will be eaten. Adult supervision is required when using ovens.

▼ Explain

- Provide the following text to the class before pupils complete the 'What I've learned' section of the worksheet.

  An insulator keeps warm things warm and cold things cold by preventing warm air from moving toward colder air. Examples of insulators are: flasks, polystyrene cups.

  In this experiment, the meringue is the insulation. The whipped egg whites are filled with tiny air bubbles. These air bubbles act as insulators and slow the hot air down as it moves through the dessert, preventing the ice-cream from becoming hot in the oven. Also, the sugar in the meringue hardens making it even more difficult for the heat to affect the ice cream.

▼ Apply

- Pupils work in groups to make a list of effective insulators located in the school and at home. The insulators may stop something from losing heat or stop something from warming up.

▼ Review and reflect

- Pupils complete the science report on page xii about the 'Ice-cream insulator' activity.

Ice-cream insulator

Task: To bake ice-cream in the oven without it melting.

▶ What to do:

1. Separate the whites from the eggs and place whites in bowl.
2. Add $1/4$ teaspoon of salt and $1/4$ teaspoon of cream of tartar and whisk until the mixture forms stiff peaks.
3. Gradually add the sugar, one spoonful at a time, then stir until thick and shiny in appearance.
4. Place baking paper on tray and place 6–8 biscuits on each tray.
5. Scoop ice-cream only on the centre of each biscuit.
6. Use a spatula to cover the ice-cream with the meringue mixture. The ice-cream and biscuit must be completely covered (no gaps!).
7. Bake at 110 °C on the bottom shelf of oven for one hour.
   Safety: Only an adult can use the oven!

You will need

- ice-cream and scoop
- biscuits
- 3 eggs
- 1 cup caster sugar
- salt
- cream of tartar
- baking paper
- bowl
- spatula
- biscuit tray
- whisk
- measuring spoons
- oven and oven gloves

▶ What happened?

Draw a diagram of a cross-section of your experiment (cut an ice-cream bake in half). Remember to add labels.

▶ What I've learned

1. How successful was your 'Ice-cream insulator' cooking experiment?
2. How could you improve it? _____
3. Explain why the ice-cream doesn't melt. (Remember to include the word 'insulator')